Heinrich W. Dove

The law of storms

Heinrich W. Dove

The law of storms

ISBN/EAN: 9783742860514

Manufactured in Europe, USA, Canada, Australia, Japa

Cover: Foto ©berggeist007 / pixelio.de

Manufactured and distributed by brebook publishing software
(www.brebook.com)

Heinrich W. Dove

The law of storms

THE

LAW OF STORMS

CONSIDERED IN CONNECTION WITH THE ORDINARY MOVEMENTS OF THE ATMOSPHERE.

BY H. W. DOVE, F.R.S.

Member of the Academies of Berlin, Moscow, Munich, St. Petersburg, Vienna, &c.

WITH DIAGRAMS AND CHARTS OF STORMS.

SECOND EDITION, ENTIRELY REVISED AND CONSIDERABLY ENLARGED.

TRANSLATED WITH THE AUTHOR'S SANCTION AND ASSISTANCE BY

ROBERT H. SCOTT, M.A.

Trinity College, Dublin: Lecturer in Mineralogy to the Royal Dublin Society.

LONDON:

LONGMAN, GREEN, LONGMAN, ROBERTS, & GREEN.

1862.

DEDICATION.

———◆———

REAR-ADMIRAL ROBERT FITZROY.

MY DEAR SIR,

Ever since the time when you conducted me over the magnificent dockyards of your country, I have received so many proofs of your kind recollection of me, that I request you to allow me to dedicate the second edition of this work to you. This I am the more desirous of doing, inasmuch as it is to you that I am indebted for the translation of the first edition, which appeared in the 'Meteorological Papers,' and for the valuable additions which were then made to it.

Yours with sincere respect,

H. W. DOVE.

AUTHOR'S PREFACE.

In this 'Second Edition' the work has been so completely remodelled and enlarged, that I have added to the original title —'The Law of Storms'— the words, 'considered in connection with the Ordinary Movements of the Atmosphere.' In fact, these general considerations form the entire of the first half of this work. In the course of the thirty-four years which have elapsed since the appearance of my first investigations on the Law of Gyration and the Rotatory Movements of Storms, I have endeavoured, as far as possible, to supply, in confirmation of my theories, the empirical proofs, of which there were at first not many to be found. I have been careful to enumerate all such confirmations as I am aware of, and must apologise if I have, through ignorance of their existence, omitted to mention any. In contrast to the custom, which is unfortunately becoming every day more universal, of bringing forward facts which have been known for more than a century as if they were new discoveries of each author, I hold it to be the duty of everyone who selects the movements of the atmosphere as the subject of his investigations to state openly what we owe to men like Dampier, Halley, Hadley, and Horsburgh, in this branch of science. The problems which are presented to us by the atmosphere are too complicated to allow of their solution off-hand; and there will ever remain questions which earlier observers have been unable to solve. To be classed with such predecessors as those I have named is the noblest recompense which we can hope for in return for our labours.

BERLIN:
June 10, 1861.

TRANSLATOR'S PREFACE.

In this translation I have, by Professor Dove's desire, embodied in the work itself the results of observations at Dorpat (p. 125) and at Toronto (p. 129), which only appeared in the Appendix to the original work. I have also added the tables for the year 1859 at Greenwich (p. 102), for Bermuda (p. 105), and for Melbourne (p. 115), which were received from the Author in the progress of the translation. Any additions which I have made in the way of notes are signed 'Trans.' The notes signed 'F.' have been transferred, by Admiral Fitzroy's permission, from the translation of the First Edition, which is contained in the third number of 'Meteorological Papers.'

The measures throughout the work have been converted to English inches, and the temperatures to their equivalents on the Fahrenheit scale. The quotations from all the works which were accessible to me in the libraries of Dublin have been taken from the originals. This has been the case with the quotations from Dampier, Halley,

CONTENTS.

———

INTRODUCTION.

PAGE

ORDINARY WINDS AND ROTATORY STORMS 3

INFLUENCE OF THE ROTATION OF THE EARTH ON THE DIRECTION OF
THE WIND 7

I. THE PERMANENT WINDS.

1. THE UNDER TRADE-WIND . . 18

2. THE UPPER RETURN TRADE-WIND 36

II. THE YEARLY PERIODICAL WINDS.

1. THE MONSOONS OF THE EAST INDIES 43
   THEIR CAUSE IS TO BE SOUGHT FOR IN THE RAREFACTION OF
   THE AIR OVER CENTRAL ASIA IN SUMMER . 50

2. THE WEST MONSOONS OF THE LINE . . . 65

3. THE LATERAL DEFLECTIONS OF THE TRADE-WIND 67

GENERAL VIEW OF THE WINDS OF THE TORRID ZONE . 73

III. THE CHANGEABLE WINDS.

GENERAL CONSIDERATIONS 77

1. THE TWO CURRENTS, EQUATORIAL AND POLAR . . . 79

2. THE LAW OF GYRATION ON THE NORTHERN HEMISPHERE 87

3. THE LAW OF GYRATION ON THE SOUTHERN HEMISPHERE 107

LIST OF CHARTS.

+

CHART I. STORM OF DECEMBER 25, 1821   *To face page* 160

" II. WEST INDIA HURRICANE OF AUGUST, 1837 .   . 177

" III. MAURITIUS STORM OF MARCH, 1809   . *ib.*

" IV. TRACKS OF WEST INDIA HURRICANES .   *To face page* 178

" V. STORM OF JUNE 3–5, 1839, BAY OF BENGAL   "   180

" VI. TRACKS OF TYPHOONS, CHINA SEAS   "   192

Errata.

Page 38, line 3 from foot, *for* Halley *read* Hadley.
　61, „ 16 „   „   „ tension „ increase of tension.
　100, „ 16 „   „   „ 1829 „ 1729.
　128, the table for Ogdensburgh should precede the notice of Toronto.

THE LAW OF STORMS

CONSIDERED IN CONNECTION WITH THE ORDINARY
MOVEMENTS OF THE ATMOSPHERE.

INTRODUCTION.

THE AIR, if it be once set in motion, either with moderate velocity or as a storm, may preserve the direction of that motion unchanged, or else blow successively from different points of the compass. In speaking of atmospherical currents, we distinguish permanent winds from those which are variable. In the same way, ordinary language has drawn a distinction between different classes of storms : those whose direction is constant being called *gales;* those which have a rotatory motion *hurricanes,* or, as Piddington calls them, *cyclones;* and lastly, rotatory storms of smaller dimensions being termed *trombs.*

We have succeeded, by the help of the 'Law of Gyration,' in referring the Trade-winds (whose direction is never changed), the Monsoons (whose direction is changed periodically), and the so-called Variable Winds of higher latitudes, to one common general principle, which Hadley had first applied to explain the origin of the Trade-winds. It is, therefore, not unnatural for us to expect that in the more violent disturbances of the atmosphere, certain general conditions will exist, which are common to them all, as well in their origin as in their subsequent course, although the appearances which they present may exhibit considerable variety. Instead of following this natural course of inquiry, identity has been sought, not in the original conditions which produced the phenomena, but in the phenomena themselves. The consequence of this error has been that a very animated discussion has arisen; one party maintaining that all storms are cyclones, while the other asserts that the rotatory motion is never developed on a large scale.

An error which is constantly refuted, and yet as often re-appears, is the belief that the meteorological phenomena of the temperate zone are only modifications, on a smaller scale, of those which appear in the torrid zone. The real state of the case is, that the torrid zone presents the simplest form of the phenomena, whose more general characters are exhibited in higher latitudes. This assertion is found to be the case as regards the movements of the atmosphere, as well in their mean direction, as in the extraordinary increase in their intensity, which we call a storm. In the case of the former, we find that the Law of Gyration is the general rule observable in the temperate zones, while its exceptional cases, the Trade-winds and Monsoons, are peculiar to the torrid zone. In the case of the latter, we find that the existence of both gales and cyclones in the temperate zone may be demonstrated, whereas the cyclonical class of storm predominates to such an extent in the torrid zone, that the other class sinks into comparative insignificance beside it. A general theory of the wind ought to explain the reason why the primary causes, which affect the whole of the atmosphere, should produce exclusively certain definite and exceptional effects in the torrid zone.

If we inquire in what cases a vane, which is exposed to the action of air in motion, will preserve its direction unchanged, we find that they are three in number : —

1. When the air flows from all sides towards one fixed

point at which it rises, or flows away, on all sides, from one fixed point where it sinks.*

2. When the air circles about one fixed point, either from left to right, or from right to left. In this case the direction of the vane at the place of observation indicates the tangent to the whirlwind, which exists at that fixed point; whereas, in the former case, it corresponds with a radius, along which the influx or efflux of the air takes place.

3. When large masses of air move from one point on the earth's surface to another, and consequently the individual particles move in parallel rectilinear paths. In the first case which we have just discussed, the paths are diverging and converging right lines, while in the second they are concentrical circles. The third case of motion, viz., that in parallel rectilinear paths, can only take place under the following conditions : —

a. If the places from which the air starts and at which it arrives be in the same latitude, no change in the distance between them, or in the velocity of the wind, can have any effect on the vane; *i.e.*, the direction of the vane is unaltered, however the extent in longitude or the velocity of the current may vary (according to Hadley).

---

* The land and sea-breezes of small islands in hot climates, which are felt when there is no prevailing wind, are examples of this. If, during the day, the land assumes a higher temperature than the sea, the air in contact with it rises, and the cooler sea-air flows in below. The cold air sinks over the surface of the sea, as it does in the shade of a passing cloud on a hot summer's day. During the night the land loses its acquired temperature more rapidly than the sea, so that the surface of the latter becomes at last warmer than that of the land, and a counter current from the land to the sea sets in. This circular motion in a vertical plane may be compared to the revolutions of a wheel. If there be no difference of temperature there is no motion ; if there be a difference, motion ensues, first in one direction, then in the other. There are two periods of rest every day, viz., those at which the direction of the rotation changes.

b. If on the other hand the latitude, and consequently the velocity of rotation of the surface with which the air is in contact, be not constant, the vane can preserve its direction unchanged, solely on the condition that the interval between the extreme points of its course remain unchanged, and at the same time its velocity do not vary.

The cases in which the vane changes its direction are also three in number : —

1. When the point towards which the wind is moving or from whence it has started, changes its position. If (fig. 1) we suppose this point to move from SW. to NE., we find that at the centre of the circle the wind chops about, after a calm, from NE. to SW.; on the right hand side of the path of the centre it veers from ENE. through E. to SW., and on the left hand side of the path, from NNE. through W. to WSW. In each of these cases the amount of change diminishes as the distance from the centre increases. The same reasoning holds for winds which diverge from one central point, if we bear in mind that the directions of the wind in this case are diametrically opposite to those in the case last considered.

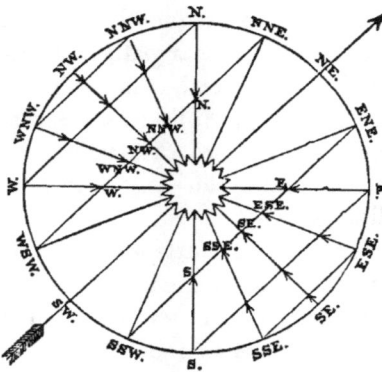

2. When the point around which the air rotates changes its position. If the mass of air, rotating from right to left, in a direction opposite to that of the hands of a watch, move from SW. to NE., the place of observation will pass successively through the points of intersec-

tion of a right line, with the concentrical circles of the
cyclone passing over it. In this case the directions of
the vane represent the
tangents to these cir-
cles drawn at the
points of intersection.
At the centre the wind
chops about from SE.
to NW.; on the right
hand side it veers from
SSE. through S. to-
wards WNW., and on
the left hand side from
ESE. through E. to-
wards NNW. The
direction in which the change of direction takes place,
on either side of the path, is the same as in the former
case ; but the directions of the wind differ by an angle
of 90° in the two cases. If the cyclone rotate from
left to right, the directions of the wind will be exactly
reversed.

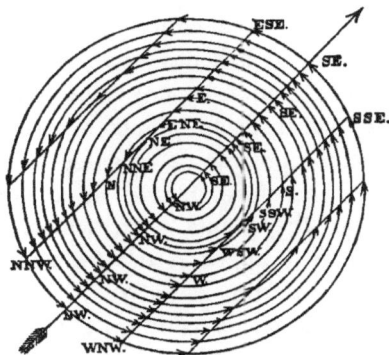

3. When, in the case of a constant wind, the distance
of the initial point or the velocity of the storm is
altered.

The effect produced by a change of position of the
initial point of the current may be perceived from the
following consideration : —

The velocity of rotation of the individual points on
the earth's surface varies as the diameter of their latitude, and therefore increases from zero at the poles to
a maximum at the equator. If the air be at rest, it
partakes of the velocity of rotation of that portion of the
globe over which it rests. Hence, if a difference of temperature, or any other cause, impart to it a tendency to

move in a direction parallel to the equator, its motion
cannot be affected by the rotation of the earth, inasmuch
as the points on the earth's surface towards which it is
moving have exactly the same velocity of rotation as
those which it has left. If, on the other hand, the air be
set in motion by any cause from the poles to the equator,
it passes from points which possess a lesser to those
which possess a greater velocity of rotation. This
air, therefore, moves at a slower rate from W. to E.;
than the points with which it comes in contact ; and
hence it appears to have a motion in the opposite direction, *i.e.* from E. to W. The deflection of the wind from
its original direction, its rate of progression being supposed unaltered, will be greater, the greater the difference
of the velocities of rotation of the point from whence it
has started, and of that where the observation is made;
or, in other words, the greater the difference of latitude
of these two places. Hence we see —

1. In the northern hemisphere winds, originally north
winds, change gradually through NE. towards E.

If we take points

$$A \; A_{,} \; A_{,,} \; A_{,,,}$$
$$B \; B_{,} \; B_{,,} \; B_{,,,}$$
$$C \; C_{,} \; C_{,,} \; C_{,,,}$$
$$D \; D_{,} \; D_{,,} \; D_{,,,}$$

so situated that $A \; B \; C \; D$ are all under the same meridian, A being the most northerly, D the most southerly
station ; and that $A \; A_{,}$, &c., lie on the same parallel of
latitude, A being the most westerly, and $A_{,,,}$ the most
easterly. If we conceive the whole body of air lying
between $A \; A_{,,,}$ and $D \; D_{,,,}$ set in motion from north to
south by any cause, we shall find that the air from $C \; C_{,,,}$
reaches the latitude $D \; D_{,,,}$ more or less as a north wind;

that from $B\,B_{///}$ as a north-east wind; while that from $A\,A_{///}$ will reach it still more as an east wind. *In this case an observer in the latitude $D\,D_{///}$ will see the vane gradually move from N. through NE. towards E.*

2. In the southern hemisphere, winds, originally south winds, change gradually through SE. towards E.

If we take, as before, points

$$d\ \ d_{,}\ \ d_{,,}\ \ d_{///}$$
$$c\ \ c_{,}\ \ c_{,,}\ \ c_{///}$$
$$b\ \ b_{,}\ \ b_{,,}\ \ b_{///}$$
$$a\ \ a_{,}\ \ a_{,,}\ \ a_{///}$$

of which those lying in the latitude $a\,a_{///}$ are the most southerly, and those in the latitude $d\,d_{///}$ the most northerly, *an observer in the latitude $d\,d_{///}$ will see the vane gradually turn from S. through SE. towards E.*

If in either hemisphere an east wind has arisen in the manner above indicated, it will preserve its motion in the latitudes $D\,D_{///}$ or $d\,d_{///}$ without being in any way affected by the rotation of the earth.*

If the cause which produced a current towards the equator continue in operation, the east wind, which has thence arisen, will interfere with and retard the current. By means of this retardation, the air will soon attain the velocity of rotation of the places with which it is in

---

* Ferrel, *The Motions of Fluids and Solids relative to the Earth's Surface: comprising Applications to the Winds and the Currents of the Ocean.* New York, 1860, p. 25. 'In whatever direction a body moves on the surface of the earth, there is a force arising from the earth's rotation which deflects it to the right in the northern hemisphere, but to the left in the southern.' This extension of the principle of Hadley's theory leads to the same conclusions as that theory in the case of all the phenomena which are here considered. If we take into consideration the components of the moving force, which result for motions in the direction of the parallels of latitude, the conclusion which has been drawn as to the change of direction of the wind holds good for all winds, including east and west.

contact, and will pass into a state of relative rest with regard to it. A constant tendency of the air to flow towards the equator will therefore reproduce precisely the phenomena which have been just considered.

. If we now take the case of the appearance of equatorial currents, after polar currents have predominated for a time — in the northern hemisphere the incipient equatorial current, as a south wind, will displace the polar current, which has become more or less easterly, in the direction E., SE., S. ; in the southern hemisphere, as a north wind, it will displace the polar current, which has become easterly like the other, in the direction E., NE., N.

Hence, on the whole, in the latitude $D\,D_{///}$, in the northern hemisphere, the observed change of the wind will be N., NE., E., SE., S.

In the latitude $d\,d_{///}$, in the southern hemisphere, it will be exactly the reverse, viz., S., SE., E., NE., N.

Air, which flows from the equator towards the poles, moves from points which possess a greater to those which possess a lesser velocity of rotation towards the east. Hence it follows that —

3. In the northern hemisphere a southerly wind in its progress gradually veers through SW. towards W.

4. In the southern hemisphere a northerly wind in its progress gradually veers through NW. towards W.

If we take points

$$
\begin{array}{llll}
D & D_{,} & D_{,,} & D_{,,,} \\
E & E_{,} & E_{,,} & E_{,,,} \\
F & F_{,} & F_{,,} & F_{,,,} \\
G & G_{,} & G_{,,} & G_{,,,}
\end{array}
$$

in the northern hemisphere, such that those in the latitude $G\,G_{///}$ are the most southerly, we shall see, as before, that if the whole body of air contained between the latitudes $D\,D_{///}$ and $G\,G_{///}$ be set in motion from the south towards the north, *an observer in the latitude $D\,D_{///}$ will receive the wind from $E\,E_{///}$ nearly as S.; that from $F\,F_{///}$ more as SW.; and that from $G\,G_{///}$ nearly as W.*

Similarly, if we take points

$$d\ \ d_{,}\ \ d_{,,}\ \ d_{,,,}$$
$$e\ \ e_{,}\ \ e_{,,}\ \ e_{,,,}$$
$$f\ \ f_{,}\ \ f_{,,}\ \ f_{,,,}$$
$$g\ \ g_{,}\ \ g_{,,}\ \ g_{,,,}$$

in the southern hemisphere, such that those in the latitude $g\,g_{///}$ are the most northerly, and those in $d\,d_{///}$ the most southerly, we shall see that if the air lying between $d\,d_{///}$ and $g\,g_{///}$ be set in motion from the north towards the south, *an observer at $d\,d_{///}$ will receive the wind from $e\,e_{///}$ nearly as N.; that from $f\,f_{///}$ more as NW.; and that from $g\,g_{///}$ nearly as W.*

A west wind, in either hemisphere, will interfere with fresh equatorial currents, and bring them to a state of relative rest. If the tendency of the air to flow towards the pole continue, the phenomena above described will be repeated until new polar currents change the west wind in the northern hemisphere, through NW. towards N., and in the southern through SW. towards S.

This gives —

For the northern hemisphere the change S., SW., W., NW.; N.

For the southern hemisphere the change N., NW., W., SW., S.

From the above considerations we derive the following laws : —

A. In the northern hemisphere, when polar and equatorial currents succeed each other, the wind veers in general in the direction S., W., N., E., S., round the compass. Exceptions to this rule are more common between S. and W. and between N. and E., than between W. and N. or between E. and S.

B. In the southern hemisphere, when polar and equatorial currents succeed each other, the wind veers in general in the direction S., E., N., W., S., round the compass. Exceptions to this rule are more common between N. and W. and between S. and E., than between W. and S. or between E. and N.

This is the phenomenon which I have termed THE LAW OF GYRATION.

The Trade-winds and the Monsoons may be considered as special cases of this law. In the torrid zone, the only atmospherical current which is felt at the surface of the ground is the polar current; and consequently a complete shift of the wind round the compass can never be observed. The deviation of the vane which is observed is proportional to the distance of the point of observation from the initial point of the current, and is only modified at the different seasons, to a certain extent, by the variation of the position of this initial point. The most obvious instance in proof of this statement is the NE. Trade-wind of the northern hemisphere. We find, however, that, owing to the preponderance of land in the northern hemisphere, the zone of maximum temperature of the globe, at which the heated air rises, does not coincide with the equator, but falls in the northern hemisphere. This zone is the district towards which the polar currents of both hemispheres flow, between the tropics ; and consequently the SE. Trade-wind crosses the line into the northern hemisphere. This wind, in the earlier portion of its course, moves over points whose velocity of rotation from W. to E. is continually increasing : on its entrance into the northern hemisphere it moves over points whose velocity of rotation is continually decreasing, and consequently the polar current is converted into an equatorial current. The wind, which had previously assumed a more easterly direction, shifts through SE. to S., and may finally become south-westerly. The configuration of the land about the Indian Ocean produces this phenomenon on the most extensive scale. In this district, the SE. Trade-wind enters the northern hemisphere as the SW. Monsoon during our summer, and the NE. Trade-wind enters the southern hemisphere as the NW. Monsoon during our winter.

The conversion of an equatorial into a polar current can only take place in consequence of the equatorial current passing over the pole, so that its deviation, which in the earlier part of its course was westerly, becomes subsequently easterly. We are as yet without observations to control our conclusions relative to this phenomenon.

If the air, which rests on any parallel of latitude, could be instantaneously transferred to another parallel, we should see the results of the difference in the velocity of rotation between this portion of the earth's surface, and the air which has been placed in contact with it, developed in their full extent. This is, however, never realised, as, in the passage of the air from one parallel to another, the surface of the ground with which it is in contact has a tendency to impart to the air its own velocity of rotation by means of friction. In the neighbourhood of the equator, the parallels of latitude increase in diameter less rapidly than in higher latitudes; and inasmuch as in the neighbourhood of the heated belt, from which the air rises, the velocity of the wind is diminished, the tendency of the earth's surface to impart its motion to the air with which it is in contact has a greater effect than elsewhere. Hence we

can easily see why the NE. Trade-wind has a less easterly
direction at its interior boundary, than in the centre of
the Trade-wind district.   In the region of the Calms, the
velocity of rotation of the air is the same as that of the
surface of the ground ; and as the change in the velocity
of the air which produces the calm cannot be a sudden
one, it must take place according to the principles just
explained.   This reasoning does not hold for the upper
counter Trade-wind (the return Trade), which preserves
its original equatorial velocity of rotation unchanged, up to
the time that it descends to the surface of the ground.   In
this case, as it is not in contact with the earth, its velocity
cannot be modified by the earth, and it is easy to see that
the mutual friction of two atmospherical currents is less
than the friction of air against water, and much less than
its friction against dry land.   From considerations like
these, we see why the Trade-winds should assume a more
definite form on the smooth and open surface of the sea
than in the interior of continents, and why the counter
Trade-wind should reach the earth with a considerable
westerly deflection, while the region of Calms is bounded
by currents which blow in directions nearly diametrically
opposite to each other.

The preceding consideration is totally independent of
the manner in which we conceive the motion of the air
between the latitudes in question to have arisen, whether
simultaneously at all points on the same meridian, or successively by suction or impulsion.   It is also immaterial
whether the currents which arise in the north and the
south are directly opposed to each other, or more or less
inclined towards each other and towards the meridian.

The influence of the velocity of the wind on these appearances will be easily understood from the preceding
explanation.   If the air move slowly, the surface over

which it is passing will impart to it more of its own velocity of rotation than if it moved more rapidly. Hence a
greater velocity of the wind will produce a greater deviation of the vane than a smaller velocity. This shows us
that a change in velocity of the wind produces a deviation
of the vane.

In the foregoing investigation, it has been shown that in
our latitudes a northerly current becomes more easterly
the longer it lasts, and therefore that a NE. wind is a N.
wind which has come from higher latitudes than the wind
which reaches us as a N. wind; and similarly, that a SW.
wind is a S. wind which has come from lower latitudes
than our S. wind. Hence we shall be prepared to admit
that a rotation of the vane can be an indication of the
existence of a permanent current. The perpetual disregard of this fact is the chief cause of the confusion which
exists, and continually appears afresh in the consideration
of questions connected with the theory of the wind. The
essential difference between the rotation of the vane produced by an ordinary atmospherical current, and that
produced either by a centripetal motion or by a whirling
motion with advancing centre, is, that in the first case, the
rotation is always in the same direction, while in the second,
it is opposite on the opposite sides of the path of the current.
If, in the northern hemisphere, we term the rotation
S. W. N. E., as is usual, 'with the sun,' or *direct*, and the
rotation in the direction S. E. N. W., 'against the sun,' or
retrograde, and in the southern hemisphere the rotation
S. E. N. W., 'with the sun,' and that S. W. N. E., 'against
the sun,' * we find that —

* Contrary to nautical usage. The expression 'with the sun' originated
in the northern hemisphere, beyond the tropic, where one looks at the sun
rising in the east, and going from left to right across the south meridian to
the west. This, of course, should be reversed in southern latitudes generally;

Permanent winds turn the vane only in a direct sense, or with the sun :

Cyclones or centripetal winds, when they have a progressive motion, produce a rotation which is direct or retrograde, according to the position of the point of observation with regard to the path of the centre :

Finally, by the interference of two permanent currents, whose directions are different, there ensue rotations of both kinds, viz., direct, when, on the west side of the wind-rose, a current is displaced by one which is more northerly; retrograde, when it is displaced by one more southerly. On the east side, on the contrary, the rotation is direct when the displacing current is more southerly than the previous one; retrograde, when it is more northerly. Currents diametrically opposed can check each other and cause a calm, producing the same phenomenon as is observed at the centre of a series of centripetal currents or of a cyclone; viz., winds diametrically opposed, separated by a calm district.

We see, therefore, that the same appearances can arise under totally different conditions, and that it is only by means of a careful investigation of all the phases of a phenomenon that we can pronounce a decision on any one that we have to do with in a particular instance. In this investigation the barometer is one of the chief means at our disposal, and we shall therefore consider its indications with especial attention.

I. Previous to entering on the discussion of the storms themselves, we shall examine more closely the motions of the atmosphere in the different zones.

We find, as has been stated above, three types of these motions :—

but seamen retain the sense of ' with the sun ' as (applicable to the coiling or lay of a rope) from left to right. Hence some confusion. — F.

1. PERMANENT WINDS : The Trade-winds of the Torrid Zone.

2. PERIODICAL WINDS : The Monsoons of the Indian Ocean.

3. CHANGEABLE WINDS : The winds of the Temperate and Frigid Zones.

I.—The Permanent Winds.—The Trade Winds.

I. THE UNDER TRADE-WIND.

IF any point in a liquid be heated more strongly than the others, currents arise in it, and the colder particles flow from all sides towards this heated point. At this point itself the fluid is at rest, owing to the compensation of the opposite motions. The most obvious illustration of this is offered by the flame of a candle when it is burning steadily.

Were the earth at rest, and the sun fixed over a point on the equator, the air from all parts of the world would flow towards this point, and we should find the conditions of the flame of the candle fulfilled. The earth, however, moves on its axis, and consequently, instead of one heated point, we have a zone of maximum temperature on which the air is at rest. This zone separates the areas of the northern and southern currents of cold air, each of which is thus forced to perform its own circuit.

If we assume that the mean breadth of this zone is invariable throughout the year, and that it moves with the sun through $23\frac{1}{2}°$ on each side of the equator, we shall find it in June under the tropic of Cancer, in January under that of Capricorn. Consequently, all places in the torrid zone will be for a part of the year in the northern Trade-wind, and for a part in the southern. These periods will be separated by intervals at which there will be no

particular direction of the wind. If the air be supposed
to flow towards the parallel of latitude which has the sun
in the zenith at the time, the wind will flow from directly
opposite quarters, within the yearly period, at all points
over which the sun passes twice in the year. The duration of the two currents will be the same at the equator.
In the northern hemisphere the northern current will last
longer, and *vice versâ* in the southern hemisphere. The
difference in the durations will increase with the distance
from the equator. At the tropics there will be only one
current interrupted by a calm at the time of solstice. We
should therefore obtain at all points in the torrid zone
' Monsoons' or winds depending on the seasons. The rotation of the earth would produce an effect on these, and
we should have the following winds at the different seasons, on the respective sides of the equator :—

TORRID ZONE.

	Northern	Southern
Spring	NE.	SE.
Summer	SW.	SE.
Autumn	NE.	SE.
Winter	NE.	NW.

In the preceding investigation we have omitted to take
notice of the fact that the magnitude of the masses of air,
which are separated by the region of calms, varies very
much at different seasons. At the winter solstice the air
of the torrid zone is included in the northern Trade-wind
circuit, and at the summer solstice in the southern circuit ;
so that it is only at the equinoxes that the quantities of
air in each of these circuits are equal. As the relation
between these masses varies with the seasons, it will necessarily interfere with the regular motion of the zone of

Trade-winds northwards and southwards. Another point which it is necessary to remember is, that the vertical Trade-wind circuit does not take place in an area which is uniform in width, but in one which tapers towards the pole. We may consider the space between two meridians as a section of an isosceles triangular prism, whose base has a higher temperature than any other part. Consequently the air which rises at the base will not descend at the vertex, but at some parallel to the base line. From this it is evident that the phenomenon of the Trade-winds will not embrace the whole of a hemisphere, but only a portion of it which is adjacent to the equator, and that the Trade-winds will have interior and exterior limits. It appears, further, that the mean position of the place of meeting of the two currents will not coincide with the equator, but will be thrown up into the northern hemisphere. The reason of this is, that the unequal distribution of dry land on the surface of the earth causes the northern hemisphere to have a higher mean temperature than the southern. The displacement of the zone of Trade-winds at the different seasons, will be influenced more by the conditions of temperature in the torrid zone than by those of the whole earth ; inasmuch as the air in districts lying at a distance from the equator is not included in the Trade-wind circuit.

We shall now proceed to examine the observations, and draw our conclusions from them.

In the ' Distribution of Heat on the Surface of the Earth *,' I have given the following tables of the mean temperature of the northern portion of the torrid zone for each month in the year.

----

* *Die Verbreitung der Wärme auf der Oberfläche der Erde*, &c. Reimer, Berlin. 1852. 4to.

Latitude	0°	10°	20°	30°
January . . .	**79·47**	77·22	70·02	58·55
February . . .	**80·15**	78·57	72·72	59·90
March	**80·60**	79·69	75·20	63·72
April	**81·27**	81·05	79·02	68·22
May	80·15	**81·27**	80·60	73·62
June	79·92	**81·05**	**81·05**	77·22
July	78·57	80·82	**81·72**	78·35
August . . .	78·80	80·82	**81·72**	80·60
September . .	79·02	**80·82**	80·60	77·45
October . . .	79·02	**80·15**	79·02	72·95
November . . .	**79·69**	**79·69**	76·32	65·97
December . . .	**79·25**	78·35	72·95	59·67
Winter	**79·69**	78·12	71·82	59·45
Spring	**80·60**	**80·60**	78·35	68·45
Summer . . .	79·02	80·82	**81·50**	78·80
Autumn . . .	79·25	**80·15**	78·57	72·05
Year	79·69	**79·92**	77·45	69·80

If we compare with this table the latitudes of the interior boundary of the Trade-winds given from observations in the Atlantic and Pacific Oceans, we find the following results :—

The northern limit of the zone of calms, i. e. the latitude at which the NE. Trade-wind disappears in the neighbourhood of the equator, is, according to D'Après :—

In January and May,	between	6° and	4° N. lat.
February,	„	3	5	„
March and April,	„	2	5	„
June	at	10		„
July, August, and September,	between	13 and 14		„

Seller* gave the following limits of the NE. Trade-wind in the year 1675, and at the same time the direction of the SE. Trade-wind which is met with after leaving the zone of the NE. Trade-wind.

In January ⎫
February ⎬ . . 4° N. lat. where the SE. and E. winds begin.
March ⎭

* Quoted in Horsburgh, *East India Directory*, vol. i. p. 26, 2nd edit.

In April . . . 5° N. Lat. where the SE. wind begins.
    May . . . 6 . . . . . . SE., a little southerly.
    June . . . 8 . . . . . . S.
    July . . . 10 . . . . . . S.
    August . . 11 . . . . . . S., a little westerly.
    September . 10 . . . . . . S.
    October . . 8 . . . . . . S., a little easterly.
    November . 6 . . . . . . SE.
    December . 5 . . . . . . SE.

From this we see clearly that the higher the SE. Trade-wind passes into north latitudes, after crossing the equator, the more is its direction altered from a south-easterly, through south, into a south-westerly. This is a necessary consequence of the diminution in the rapidity of rotation of the surface of the ground with which the air is in contact. Dampier gives the direction of the wind in the summer months between the equator and 12° N. lat., as SSE., SSW., and SW. Horsburgh (*India Directory*, vol. i. p. 16) says, that the SE. Trade-wind, at its northern limit, has a nearly southerly direction, especially in July, August, and September. The same is true of other months. If you go further southward, the wind is more south-east. Basil Hall (*Fragments of Voyages and Travels*, second series, vol. i. p. 189) expresses himself quite as decidedly about the southerly winds which prevail at the northern limit of the SE. Trade-wind. He asserts that the easterly direction of the Trade-winds at their meeting, which is usually given on charts, is absolutely false, but is so prevalent an opinion that young officers are usually much astonished to find themselves convinced of its error by their own experience.

Horsburgh (vol. i. p. 25) gives the following table of the limits of the Trade-winds between the 18th and 26th degrees of W. long. It is taken from the observations of

149 ships, passing from the NE. Trade into the SE., and of 88 passing from the SE. into the NE. The latitude is north.

	Southern limit of the NE. Trade.		Northern limit of the SE. Trade.		Breadth of the Intermediate zone.	
January . . .	5°	45'	2°	45'	3°	0'
February . .	6	...	1	15	4	45
March . . .	5	8	1	15	3	45
April	5	45	1	15	4	30
May	6	30	2	45	3	45
June	9	...	3	...	6	...
July	12	...	3	30	8	30
August . . .	13	...	3	15	9	45
September . .	11	45	3	...	8	45
October . . .	10	...	3	...	7	...
November . .	8	...	3	45	4	15
December . .	5	30	3	15	2	15

and hence for the different seasons :—

Winter . . .	5°	45'	2°	25'	3°	20'
Spring	5	47	1	45	4	2
Summer . . .	11	20	3	15	8	5
Autumn . . .	9	55	3	15	6	40
Year	8°	12'	2°	20'	5°	52'

Kerhallet gives the following table of the extent of the Trade-winds in the Pacific Ocean, compiled from the observations of 92 ships, (*Considérations générales sur l'Océan Pacifique*, 1856, p. 4) : —

	POLAR LIMIT				EQUATORIAL LIMIT				Breadth of the region of calms.	
	of the NE. Trade, Lat. N.		of the SE. Trade, Lat. S.		of the NE. Trade, Lat. N.		of the SE. Trade, Lat. N.			
January . .	21°	0'	33°	25'	6°	30'	5°	0'	3°	30'
February . .	28	28	28	51	4	1	2	0	2	1
March . . .	29	0	31	10	8	15	5	50	2	25
April . . .	30	0	27	25	4	45	2	0	2	45
May . . .	29	5	28	24	7	52	3	36	4	16
June . . .	27	41	25	0	9	58	2	30	7	28
July . . .	31	43	25	28	12	5	5	4	7	1
August . .	29	30	24	18	15	0	2	30	12	30
September .	24	20	24	51	13	56	8	11	5	45
October . .	26	6	23	27	12	20	3	32	8	48
November .	25	9	28	39	...		...		...	
December . .	24	0	22	30	5	12	1	56	3	16

From these tables we see that the variation in position of the zone of Calms takes place within narrower limits than that of the zone of maximum temperature, both as regards its approach to the equator as its motion away from it. We must, however, remember that the mean temperature at any latitude exerts an influence on that of the whole earth, while the position of the zone of Calms is only affected by that of the zone of Trade-winds.

In the summer the south-east Trade-wind crosses the line and extends, as south-west Monsoon, up to the foot of the Himalayas. If we consider this to be the interior limit of the southern Trade-wind, it is evident that in the interior of Africa it must extend as far as the coast of Guinea. In fact, the inner northern limit of the southern Trade-wind may thus be shown, by considering the south-west Monsoon of the Indian Ocean as a modification of that wind, to lie much further to the north of the equator than would appear to be the case from the consideration of the Trade-wind itself. The correspondence between the distribution of temperature and of the Trade-wind areas is consequently much closer than it would appear to be if we were to neglect the Monsoons. In the same way the same causes affect the approach of the interior southern limit of the NE. Trade to the equator, inasmuch as that wind in the Indian Ocean, during the winter months of the northern hemisphere, does not only reach the equator, but crosses it and appears in the southern hemisphere as north-west Monsoon. We are not yet in a position to determine the extent to which this takes place.

Maury gives the following determinations of the extreme limits of the Trade-wind in the Atlantic Ocean. In summer this wind commences near the Azores, in winter to the south of the Canary Islands :—

Longitude, W.	Latitude of commencement of NE. Trade in			
	Winter.	Spring.	Summer.	Autumn.
70°	28°	28° ·7	29° ·3	29°
65	26 ·3	28 ·0	29 ·3	28 ·3
60	24	24 ·3	27 ·3	28 ·3
55	22	22 ·7	24 ·7	25
50	21	23 ·7	28 ·3	23 ·7
45	23	24 ·7	31 ·3	28 ·7
40	27 ·7	29 ·7	30 ·7	29 ·3
35	26	27 ·3	30 ·7	25 ·7
30	24 ·3	28 ·7	29 ·7	26 ·7
25	25 ·3	24 ·7	31 ·3	26 ·3
20	24 ·3	28 ·3	28 ·7	27
15	29	31	32	31 ·3
10		31 ·3	34 ·7	32

EXTENT OF THE TRADE IN GEOGRAPHICAL MILES.

Longitude, W.	Winter.	Spring.	Summer.	Autumn.	Year.
50°	274·5	280·5	285	205·5	261
45	304·5	319·5	340·5	295·5	315
40	375	415·5	325·5	300	354
35	355·5	400·5	325·5	244·5	332
30	334·5	405	289·5	270	325
25	330	330	310 5	255	306
20	304·5	364·5	264	250·5	296
15	346·5	379·5			

A balloon allowed to drift freely before the wind, although the direction of its motion may change frequently, will nevertheless be found, after the lapse of a certain period, at a definite distance in a determinate direction from the point of commencement of its motion. Lambert's formula gives this direction and the extent of the motion, considering as 100 the distance which it would have travelled if its motion during the whole period had been exclusively in this direction. Coffin (*Winds of the Northern Hemisphere*) has determined in this way the direction and intensity of the NE. Tradewind in the Atlantic Ocean.

Coffin (*loc. cit.* p. 158) gives the following table of the directions : —

REGION OF CALMS.

Latitude . . . . Longitude . . . . Days	0°— 5° 3005	5°— 10° 10°— 55° 3339
January	S 81° 46′ E	N 47° 5′ E
February	N 83 31 E	N 44 56 E
March	N 63 13 E	N 45 3 E
April	N 52 18 E	N 44 50 E
May	S 89 59 E	N 55 38 E
June	S 47 45 E	S 59 1 E
July	S 37 17 E	S 7 1 E
August	S 20 52 E	S 4 59 W
September . . .	S 20 15 E	S 8 26 W
October	S 38 0 E	S 38 2 E
November . . .	S 58 28 E	S 82 15 E
December	S 68 23 E	N 60 25 E
Year	S 60° 2′ E	N 80° 32′ E

EASTERN SIDE OF THE TRADE-WIND ZONE.

Latitude . Longitude. Days .	10°— 15° 15°— 45° 1850	15°— 20° 15°— 45° 1332	20°— 25° 15°— 45° 1334	25°— 30° 15°— 45° 1622	30°— 35° 15°— 45° 1749
January .	N 55° 30′E	N 50° 42′E	N 64° 9′E	N 78° 26′E	S 46° 8′E
February .	N 54 41 E	N 46 48 E	N 56 50 E	N 43 35 E	S 2 47 E
March . .	N 55 51 E	N 49 29 E	N 26 2 E	N 80 19 E	S 27 53 E
April . .	N 56 44 E	N 49 28 E	N 43 0 E	N 79 39 E	S 1 29 W
May . . .	N 49 14 E	N 43 50 E	N 45 34 E	N 67 39 E	N 88 32 E
June . .	N 55 0 E	N 42 8 E	N 48 49 E	N 42 48 E	N 30 9 W
July . .	N 57 2 E	N 41 26 E	N 37 45 E	N 44 35 E	N 32 35 E
August . .	N 49 18 E	N 40 49 E	N 42 1 E	N 53 11 E	S 76 13 E
September	N 46 6 E	N 54 14 E	N 51 8 E	N 62 36 E	N 14 40 E
October .	N 69 20 E	N 54 50 E	N 57 58 E	N 73 31 E	N 45 21 E
November.	N 68 54 E	N 60 50 E	N 67 70 E	N 78 50 E	S 21 58 E
December .	N 61 33 E	N 58 5 E	N 65 9 E	N 70 27 E	S 42 25 E
Year . .	N 57° 25′E	N 49° 1′E	N 55° 20′E	N 62° 53′E	S 44° 27′E

WESTERN SIDE OF THE TRADE-WIND ZONE.

Latitude . Longitude Days . .	10°—15° 45°—75° 662		15°—20° 45°—80° 1190		20°—25° 45°—80° 1573		25°— 30° 45°— 80° 2906		30°— 35° 45°— 75° 2564	
January .	N 55°	0'E	N 64°	21'E	N 65°	29'E	N 42°	50'E	S 80°	10'W
February .	N 52	12 E	N 58	25 E	N 75	53 E	N 55	7 E	S 79	16 W
March . .	N 58	14 E	N 67	21 E	N 72	33 E	N 74	23 E	S 73	19 W
April . . .	N 59	59 E	N 77	27 E	N 82	4 E	N 78	31 E	S 49	3 W
May. . . .	N 63	8 E	N 68	21 E	N 80	1 E	S 63	52 E	S 62	43 E
June . . .	N 51	50 E	N 60	20 E	N 80	42 E	S 43	17 E	S 22	27 W
July. . . .	N 56	49 E	N 62	25 E	N 78	24 E	S 67	2 E	S 8	41 E
August . .	N 65	14 E	N 70	38 E	N 72	6 E	S 74	51 E	S 7	11 E
September	N 82	29 E	N 83	32 E	N 83	0 E	S 81	43 E	S 49	8 E
October. .	N 73	52 E	N 83	49 E	S 68	49 E	S 69	14 E	N 85	7 E
November	N 57	37 E	N 75	48 E	N 79	1 E	N 66	16 E	S 84	32 W
December	N 45	38 E	N 61	5 E	N 69	52 E	N 70	26 E	N 81	21 W
Year . . .	N 59°	55'E	N 68°	34'E	N 79°	23'E	N 79°	4'E	S 31°	35'W

The following are the values of the intensity, 100 being taken to represent the intensity if the wind had blown in the determined direction without interruption: —

EASTERN SIDE OF THE TRADE-WIND ZONE.

Latitude.	0°—5°	5°—10°	10°—15°	15°—20°	20°—25°	25°—30°	30°—35°
January .	53	65	85	75	38	19	12
February	54	72	81	71	53	11	25
March . .	52	74	89	68	21	3	9
April . .	56	82	88	80	51	8	31
May. . .	48	69	90	81	67	8	8
June . .	69	33	75	90	74	35	1
July. . .	82	45	42	99	85	57	22
August .	84	71	17	75	84	31	12
September	79	58	23	76	71	33	13
October .	72	30	55	67	50	27	8
November	80	55	78	78	53	20	29
December	56	52	78	75	59	38	26
Year . .	55	34	66	77·5	58	26	10

WESTERN SIDE OF THE TRADE-WIND ZONE.

Latitude .	10°— 15°	15°— 20°	20°— 25°	25°— 30°	30°— 35°
January	87	78	35	15	16
February	90	86	51	10	30
March	87	75	38	3	21
April	83	64	46	25	11
May	89	84	65	43	14
June	96	95	65	44	29
July	89	87	81	57	35
August	85	80	76	47	19
September . . .	55	73	54	35	19
October	67	70	55	39	18
November	89	72	52	45	10
December	74	73	57	2	29
Year	82	77	55	28	11

From these tables we obtain the following conclusions : — Between the equator and the parallel of 5° N. the wind is southerly throughout the year, with the exception of the months of February and March. Between 5° and 10° it is southerly in summer and autumn ; northerly in winter and spring. Between 10° and 25° it is north-easterly during all the months. Between 25° and 30° this direction holds only on the eastern side of the Atlantic Ocean, as on its western side the wind is southerly during the summer months. From this we see that the exterior limit of the NE. Trade is a line which rises gradually from America to the north of Africa, so that a ship, on its passage from the temperate into the torrid zone, enters the Trade-wind on the eastern side of the Atlantic Ocean earlier than it would on the western side. Between 30° and 35°, southerly winds are prevalent over the whole ocean, but yet in such a way that their mean direction is more nearly SW. on the American side, and SE. on the European.

The way in which this mean direction is compounded of the individual observed winds, is apparent from the

following table taken from Maury, (*The Winds at Sea,
their Mean Direction, and Annual Average Direction
from each of the Four Quarters*):—

NORTH ATLANTIC OCEAN.

Lat. N.	NE.	SE.	SW.	NW.	Calm.
0°— 5° .	85	192	49	11	28
5 — 10	136	91	86	18	34
10 — 15	244	60	24	19	18
15 — 20	244	89	10	13	9
20 — 25	203	96	25	25	16
25 — 30	127	99	67	51	21
30 — 35	86	88	101	73	17
35 — 40	74	65	126	86	14
40 — 45	58	68	123	100	16
45 — 50	52	57	136	107	12
50 — 55	52	85	128	86	14
55 — 60	49	51	164	95	6

The figures in the first column of the table indicate the
number of days on which the direction of the wind was
between N. and E. ; in the second the number on which
it was between E. and S., &c. The appearance of the
SE. Trade in the northern hemisphere is clearly indicated,
as well as the predominant south-westerly direction in the
temperate zone, and the prevalence of calms on the inner
borders of both Trade-winds. The following is a similar
table for the

SOUTH ATLANTIC OCEAN.

Lat. S.	NE.	SE.	SW.	NW.	Calm.
0°— 5°	26	314	17	4	4
5 — 10	24	329	10	2	...
10 — 15	58	295	8	2	2
15 — 20	89	244	14	12	6
20 — 25	123	157	37	39	9
25 — 30	109	124	62	62	8
30 — 35	67	108	91	89	10
35 — 40	52	55	114	135	9
40 — 45	53	35	125	142	10
45 — 50	54	24	123	155	9
50 — 55	65	19	129	146	6
55 — 60	48	18	121	167	11

We see that the phenomena of the SE. Trade are even more clearly marked than those of the NE. Trade. The reason of this is evidently to be sought for in the fact that the predominance of water surface on the southern hemisphere secures a freer and more regular motion to the atmosphere. Another circumstance, the reasons of which will be hereafter indicated, exerts an influence, viz. that the number of cyclones in the northern portion of the torrid zone is greater than in the southern. If we count the wind from N. through E. to SW. and N. from 0° to 360°, we find the direction in the different quadrants to be nearly as follows : —

Latitude.	Northern Hemisphere.				Southern Hemisphere.			
	1	2	3	4	1	2	3	4
0°— 5°	52°	139°	204°	327°	60°	127°	213°	316°
5 — 10	53	138	210	288	69	133	211	317
10 — 15	54	115	223	318	63	130	206	327
15 — 20	55	110	218	327	55	128	209	332
20 — 25	50	125	219	322	45	133	217	327
25 — 30	49	128	219	317	44	139	218	334
30 — 35	44	138	222	314	39	139	222	317
35 — 40	42	141	225	312	38	138	227	314
40 — 45	43	141	226	310	35	142	228	315
45 — 50	45	139	228	309	27	135	234	316
50 — 55	49	139	228	308	30	141	237	315
55 — 60	35	140	227	311	36	144	239	310

For the Pacific Ocean the following table is given : —

PACIFIC OCEAN.

Lat. N.	NE.	SE.	SW.	NW.	Calm.
0°— 5°	49°	206°	89°	12°	9°
5 — 10	134	134	54	20	14
10 — 15	266	43	19	27	10
15 — 20	243	49	15	45	13
20 — 25	212	66	25	52	10
25 — 30	142	99	60	50	14
30 — 35	96	109	92	56	12
35 — 40	93	70	96	93	13
40 — 45	66	68	113	102	16
45 — 50	62	74	110	107	12
50 — 55	56	72	123	101	13
55 — 60	58	92	119	81	15

PACIFIC OCEAN.

Lat. S.	NE.	SE.	SW.	NW.	Calm.
0°— 5°	76°	229°	27°	25°	8°
5 — 10	62	243	25	25	10
10 — 15	96	219	14	25	11
15 — 20	89	216	18	29	13
20 — 25	91	195	37	29	13
25 — 30	80	154	77	41	13
30 — 35	63	103	112	74	13
35 — 40	63	54	129	106	13
40 — 45	56	35	125	141	8
45 — 50	49	32	129	147	8
50 — 55	41	36	118	164	6
55 — 60	86	30	95	186	8

In the West Indian Sea —

Lat. N.	NE.	SE.	SW.	NW.	Calm.
10°— 15°	260°	82°	6°	5°	12°
15 — 20	209	120	10	12	14
20 — 25	191	103	19	30	22
25 — 30	128	106	53	58	20

If we compare the numerical values for the two oceans we can easily understand that while, on the one hand, the Spanish seamen called the tropical portion of the Atlantic Ocean *el golfo de las Damas* (the Ladies' Sea), because it was so easy of navigation that a girl might take the helm; on the other hand Varenius should say that the sailors on leaving Acapulco might go to sleep, without minding the helm, as the wind was certain to take them to the Philippine Islands without altering their course.

Maury gives the following table of the exact directions in the four quadrants : —

PACIFIC OCEAN.

Latitude.	Northern Hemisphere.				Southern Hemisphere.			
	1	2	3	4	1	2	3	4
0°— 5°	57°	136°	210°	318°	59°	130°	216°	321°
5 — 10	54	136	213	314	57	120	229	311
10 — 15	51	120	227	320	55	126	225	322
15 — 20	51	117	227	322	58	126	218	319
20 — 25	53	118	225	322	58	128	221	321
25 — 30	53	127	223	318	54	131	219	316
30 — 35	50	132	212	315	48	139	219	311
35 — 40	42	139	222	316	38	141	223	314
40 — 45	47	133	224	311	35	140	228	313
45 — 50	43	134	227	310	31	135	233	312
50 — 55	44	137	228	308	36	141	236	307
55 — 60	47	137	227	307	38	137	236	311

The results obtained would be rendered more easy of comprehension if they had been calculated according to Lambert's method, in which the observed directions of the wind in the four quadrants are all projected on two directions at right angles to each other, viz. E. and W., and N. and S. The method which appears best adapted for seafaring men, is that of a graphical representation of the winds observed at various localities on the surface of the sea. This method has been followed with great success in the 'Wind Charts' of the Board of Trade, &c., published by Admiral Fitzroy in 1856, and in the 'Wind Charts of the South Atlantic Ocean,' and those ' of the Western and Eastern Portions of the Indian Ocean ' for each month of the year.* In these cases, however, it

* *Windkaart van den Zuider Atlant. Oceaan — van het Westelijk Deel —* and *van het Oostelijk Deel der Indische Zee.* Published in the 2nd and 3rd volumes of the *Uitkomsten van Wetenshap en Ervaring aangaande Winden en Zeestroomingen in Sommige Gedeelten van den Oceaan. Uitgegeven door het Kon. Ned. Meteor. Instituut.* Utrecht, 1859. (Results of Science and Experiment respecting the Winds and Marine Currents in certain Parts of the Ocean. Published by the Royal Meteorological Institute of the Netherlands. Utrecht, 1859.)

would be necessary to give the numerical values from which the lines have been constructed, in order that the mean direction might be calculated with accuracy from the results.

These numerical values are given in the 'Monthly Sailing Directions for the Voyage from Java to the British Channel'* for the Atlantic Ocean, from which work I have taken the following tables. In the northern hemisphere I have combined together the three months corresponding to the yearly meteorological seasons; while for the southern I have only given the total for the whole year, inasmuch as the variation in that hemisphere in the course of the year is very slight.

NORTHERN HEMISPHERE.

WINTER.

Latitude . . Long. W. .	30°—25° 25°—45°	25°—20° 25°—45°	20°—15° 25°—45°	15°—10° 20°—40°	10°— 5° 15°—35°	5°— 0° 15°—35°
N.	63	34	25	6	39	59
NNE.	143	84	59	82	325	172
NE.	208	271	437	620	648	256
ENE.	231	288	292	146	147	87
E.	153	84	56	12	59	146
ESE.	80	59	7	...	13	114
SE.	53	31	1	...	19	273
SSE.	40	17	...	...	19	291
S.	59	17	...	...	22	120
SSW.	36	11	...	...	11	36
SW.	33	3	...	...	9	9
WSW.	3	...	...	...	4	8
W.	19	2	...	...	10	20
WNW.	16	8	...	...	2	26
NW.	36	16	2	...	3	27
NNW.	35	13	14	...	3	23

* *Maandelijksche Zeilandwijzingen van Java naar het kanaal.* Utrecht, 1859.

SPRING.

Latitude . . Long. W. .	30°—25° 25°—45°	25°—20° 25°—45°	20°—15° 25°—45°	15°—10° 20°—40°	10°— 5° 15°—35°	5°— 0° 15°—35°
N.	34	40	10	23	186	72
NNE.	41	86	36	186	655	229
NE.	208	244	476	588	561	230
ENE.	241	381	422	127	67	101
E.	223	160	85	9	51	187
ESE.	126	62	19	...	16	215
SE.	73	19	3	...	16	157
SSE.	44	9	6	...	14	148
S.	32	12	2	...	28	103
SSW.	41	23	2	...	15	28
SW.	46	18	6	...	16	27
WSW.	21	12	3	...	5	12
W.	20	15	8	...	14	32 ·
WNW.	24	8	2	...	12	11
NW.	44	11	6	1	46	25
NNW.	31	16	4	6	96	34

SUMMER.

Latitude . . Long. W. .	30°—25° 25°—45°	25°—20° 25°—45°	20°—15° 25°—45°	15°—10° 20°—40°	10°— 5° 15°—35°	5°→ 0° 15°—35°
N.	16	1	12	159	31	2
NNE.	69	22	86	284	73	2
NE.	279	371	475	373	86	2
ENE.	279	298	174	111	32	6
E.	103	53	16	28	22	37
ESE.	44	13	...	23	16	112
SE.	25	4	...	25	27	191
SSE.	5	...	...	24	92	279
S.	7	...	1	42	139	122
SSW.	...	...	...	23	127	69
SW.	4	...	...	46	113	11
WSW.	2	...	...	48	68	3
W.	19	...	...	86	76	...
WNW	...	...	1	53	35	4
NW.	9	3	1	45	52	2
NNW.	6	...	...	33	10	3

AUTUMN.

Latitude . Long. W. .	30°—25° 25°—45°	25°—20° 25°—45°	20°—15° 25°—45°	15°—10° 20°—40°	10°— 5° 15°—35°	5°— 0° 15°—35°
N.	62	68	14	51	30	1
NNE.	111	131	123	238	122	3
NE.	183	284	423	431	216	5
ENE.	220	292	179	142	118	6
E.	110	132	76	40	81	20
ESE.	69	37	11	20	42	58
SE.	32	6	7	13	62	163
SSE.	20	...	2	15	84	348
S.	25	5	...	35	114	187
SSW.	32	...	...	11	58	92
SW.	30	...	3	52	66	23
WSW.	13	...	2	42	50	18
W.	11	...	...	18	68	5
WNW.	19	2	1	24	56	6
NW.	39	10	...	23	32	1
NNW.	37	30	...	39	14	4

SOUTHERN HEMISPHERE.

YEAR.

Latitude Longitude	0°— 5° 15°—25°W	5°—10° 10°—15°W	10°—15° 5°—15°W	15°—20° 0°—10°W	20°—25° 5°W—10°E	25°—30° 0°—15°E	30°—36° 10°—20°E
N.	...	...	...	...	11	34	151
NNE.	...	...	...	...	7	11	32
NE.	16	7	...	24	27	20	119
ENE.	58	45	66	30	19	10	79
E.	282	173	223	204	155	45	173
ESE.	640	277	1060	818	506	175	309
SE.	1950	2530	2478	2474	2488	1722	1147
SSE.	567	358	279	507	658	821	762
S.	147	39	50	149	383	587	807
SSW.	21	...	...	34	97	274	574
SW.	11	...	...	8	102	360	688
WSW.	...	...	...	...	23	230	437
W.	...	...	...	...	29	211	576
WNW.	...	...	...	...	12	105	418
NW.	...	...	...	...	22	156	549
NNW.	...	...	...	...	7	19	151

The greater constancy of the SE. Trade is very clearly seen from this, as well as its passage over the equator into

the northern hemisphere. On the whole, the results
obtained by Dutch vessels confirm the calculations made
by Coffin, from the data collected by Maury.

2. The Upper Return-Trade-Wind.

In the preceding section we have seen that the air flows
towards the equator from the poles of both hemispheres.
In consequence of this tendency it would accumulate in
the tropical regions, and its pressure, as indicated by the
barometer, would continually increase, were it not that an
efflux of the air in the opposite direction takes place in
the upper strata of the atmosphere.   A fact which bears
visible testimony to the existence of this current is the
motion of clouds, at a great height, against the Trade-
wind, which has been observed from the surface of the
earth, and expressly mentioned by many navigators.   Of
these we shall only cite Basil Hall and Paludan.   Fendler,
at Tovar in Venezuela, has proved the existence of the
current numerically by observations carried on for a long
series of years.   The level at which this return current
commences is so high, that it has not been ascertained with
certainty by the ascent of the highest peaks of the Cor-
dilleras in the vicinity of the region of Calms ; but yet the
fact of its existence in that region has been clearly
demonstrated.   In the night of the 30th of April and 1st
of May, 1812, explosions, as if of heavy ordnance, were
heard at Barbadoes, so that the garrison of Fort St. Anne
was kept under arms.   At break of day on the 1st of
May, the eastern part of the horizon was clear, while the
remainder of the sky was enveloped in a black cloud.
The darkness soon extended over the whole sky to such a
degree that the place of the windows could not be seen
in the apartments, while the trees gave way under the

load of ashes which fell on them. Where did these ashes come from? In the months of April and May the Trade-wind is at its height, so that it would have been natural to conclude that they had come from the volcanoes of the Azores. The true source of the ashes was the volcano of Morne Garou, in the island of St. Vincent, which lies 100 miles west from Barbadoes, and from which it is impossible to reach Barbadoes by sea without taking a very circuitous route, in consequence of the direction in which the Trade-wind blows. The ashes from the volcano had been thrown by the violence of the eruption through the under Trade-wind into the upper counter-current. Again, on the 20th of January, 1835, the whole isthmus of Central America was shaken by an earthquake which accompanied the eruption of the volcano of Coseguina, on the Lake of Nicaragua. The violence of the eruption was enormous, so that the sounds were heard at Sta Fé de Bogotá, at a distance of 1,000 miles; while the cloud of ashes was so dense that Union, a seaport town on the west coast of the Bay of Conchagua, was enveloped in total darkness for forty-three hours. Ashes fell also at Kingston, and at other places in Jamaica, so that the inhabitants were able to learn that the explosions which they had heard had not been those of cannon. These ashes could not have been carried to Jamaica, except by the counter-trade-wind, as that island lies to the north-east of Nicaragua.

The fact that ashes from low volcanoes, like Morne Garou and Coseguina, reached the upper current, proves that the eruptions must have been of extraordinary violence. This was the case with Morne Garou, as the eruption to which we have referred formed part of a series of stupendous volcanic phenomena, and marked the finale of the disturbance. In June and July, 1811, Sabrina

Island rose out of the sea, near St. Michael, one of the Azores. The island was raised to a height of 300 feet above the level of the sea, which is 150 feet deep at the place, and it was an English mile in circumference. This event was followed by the continual earthquakes which were felt for months in Arkansas and Ohio, and finally by the total destruction of the city of Caraccas on the 26th of March, 1812. It was not until May that the volcanic energies, which had been so long struggling to get free, burst open the vent of Morne Garou, which had been closed for a century. The noise of the eruption was heard at Rio Apure, which is about as far from the volcano as Naples is from Paris.

Halley was the first to assert the existence of an upper current in the opposite direction to the Trade-wind.* He says, ' The north-east Trade-wind below will be attended with a south-westerly above, and the south-easterly with a north-westerly above : that this is more than a bare conjecture, the almost instantaneous change of the wind to the opposite point, which is frequently found on passing the limits of the Trade-winds, seems to assure us.'

Halley accordingly considered the south-west wind at the outer limit of the north-east Trade, and the north-west wind at the outer limit of the south-east Trade to be the upper Return-Trade-wind reaching the surface of the earth. The only reason he gives for it is, that it is produced ' by a kind of circulation.' The phenomenon is, however, a necessary consequence of Halley's theory, who says †, ' The north-east and south-east winds between the tropics must be compensated by as much south-west

---

* ' An Historical Account of the Trade-winds and Monsoons observable in the Seas between and near the Tropic, with an Attempt to assign the Physical Cause of the said Wind.'— *Phil. Trans.* 1686, p. 167.

† ' The Cause of the General Trade-wind.'— *Phil. Trans.* 1735, p. 62.

and north-west winds in other parts, and generally all winds from any one quarter must be compensated by a contrary wind somewhere or other; otherwise some change must be produced in the motion of the earth round its axis.'

Leopold von Buch was the first to point out the effect which this descent of the Return-Trade-wind has on the winds of the temperate zone. This he has done in his remarks on the climate of the Canary Islands.*

It is very remarkable and instructive, as well as of the greatest importance for the science of meteorology, to observe the mode in which this NE. Trade is displaced by the SW. winds towards winter. These winds do not commence to the southward and move up northwards, but are felt first on the coast of Portugal, then at Madeira, and afterwards at Teneriffe and the Canaries. They descend as gradually from the upper strata of the atmosphere as they come down from the northward. They had existed continually at this high level, even throughout the summer, when the NE. Trade-wind was blowing with the greatest violence at the sea-level; for the peak of Teneriffe is high enough to reach the upper current even at midsummer. It is hard to find any account of an ascent of the peak in which the strong west wind which had been met with on the summit is not mentioned. Humboldt ascended the peak on the 21st of June; when he reached the edge of the crater he could scarcely keep his feet, such was the violence of the west wind (*Relat.* i. p. 132). If such a wind had been felt at Santa Cruz or Orotava at that season, the inhabitants would have been quite as much astonished as those of Barbadoes were at the ashes which fell there. I found a similar west wind, although not so high, at the summit of the peak on the 19th of May. George Glass, an attentive and accurate observer, who as a seaman had closely studied the winds of the Canary Islands for many years, says in his

* *Physikalische Beschreibung der Canarischen Inseln* (Physical Description of the Canary Islands), 1825, p. 67.

'History of the Canary Islands' (p. 251), that 'a strong west-erly wind is constantly blowing at the highest points of these islands, during the prevalence of the NE. wind below. This,' he adds, 'I believe to be the case in all parts of the world where the Trade-winds are felt. I do not venture to explain this phenomenon, but so it is at the top of the Peak of Teneriffe, and on the mountains of some others of these islands.' Glass knew the islands too well not to speak from his own experience on this subject.

These winds descend slowly down the sides of the mountains from the higher strata of the atmosphere. This is clearly to be seen by the clouds from the south, which have enveloped the top of the peak ever since October. They appear lower and lower, and at last rest on the crest of the mountains lying between Orotava and the south coast, which are about 6,000 feet high, and break up there in fearful thunder-storms. Perhaps a week or more elapses after this before they are felt at the sea-level. There they stay for months. Rain falls only on the slopes of the mountains, and the peak covers itself with snow. Can one help believing that the west wind which sailors look for, on the summer voyage from Teneriffe to England, in the neighbourhood and at the level of the Azores, and which they usually find there — that the nearly invariable west or south-west wind, which makes sailors call the voyage from New York or Philadelphia to England *down-hill*, and that from England back again *up-hill*, is not, as well as the west wind on the summit of the peak, the upper equatorial current, which descends here to the level of the sea? It would follow from this that the upper equatorial current does not reach the pole, at least over the Atlantic Ocean.

We are indebted to Piazzi Smith * for more accurate data as to the boundaries of the two currents. The vertical depth of the under NE. Trade was found to be 9,000 feet. The cloud stratum did not lie, during the preva-

* 'An Astronomical Experiment on the Peak of Teneriffe.'— *Phil. Trans.* 1859, p. 527.

lence of the NE. wind below, between that current and the SW. current above, but was to be found nearly in the centre of the under current, at a height of about 4,000 or 5,000 feet above the sea-level.

The fact that Goodrich found, in April, a SW. wind at the top of Mouna Kaa, while the NE. Trade-wind was blowing at the lower levels of the island of Hawaii, proves that the phenomena observed at the Canary Islands are true for all places situated at the outer edge of the Trade-wind zone.

If we were able to determine by actual experiments on a series of lofty peaks the direction of the wind at various heights, we should be in a position to gain a more accurate insight into the ratio between the volumes of the direct and counter currents. The barometer gives us the total pressure produced by the conjoint action of both currents, and by means of this instrument we can see very clearly that the atmospheric pressure at the interior edges of the Trade-wind zones, where the air rises, is materially less than at their outer edges, where the upper Trade-wind descends. A. von Humboldt was the first to draw attention to the diminution of pressure in the vicinity of the equator, and L. von Buch to the relatively increased pressure in the vicinity of the Canary Islands; but A. Erman and Herschel were the first to prove that the passage of the one condition into the other was gradual. It is very clearly perceptible in the new tables of observations made by the Dutch marine. We may also observe in these tables the close accordance between the variable place of ascent and the district of maximum temperature, pointed out above (page 21). The following table gives, in decimals of an inch, the differences from the mean height of the barometer observed in the Atlantic Ocean, from lat. 35° N. to 36° S., and also the limits of the zone of Calms.

ATMOSPHERICAL PRESSURE IN THE TRADE-WIND ZONE
ATLANTIC OCEAN.

Latitude	Winter	Spring	Summer	Autumn
N 35°— 30°	+ 0·192	+ 0·122	+ 0·171	+ 0·083
30 — 25	+ 0·173	+ 0·145	+ 0·134	+ 0·078
25 — 20	+ 0·067	+ 0·087	+ 0·050	− 0·002
20 — 15	+ 0 007	+ 0·008	− 0·048	− 0·049
15 — 10	− 0·062	− 0·042	− 0·103	− 0·093
10 — 5	− 0·104	− 0·107	− 0·120	− 0·112
5 — 0	− 0·119	− 0·133	− 0·121	− 0·093
S 0 — 5	− 0·117	− 0·111	− 0·087	− 0·069
5 — 10	− 0·072	− 0·076	− 0·056	− 0·024
10 — 15	− 0·033	− 0·029	− 0·004	+ 0·027
15 — 20	+ 0·007	+ 0·018	+ 0·045	+ 0·049
20 — 25	+ 0·042	+ 0·050	+ 0·077	+ 0·124
25 — 30	+ 0·050	+ 0·050	+ 0·073	+ 0·085
30 — 36	− 0·030	+ 0·029	+ 0·028	+ 0·046
Mean height .	30·020 in.	30·025	30·057	30·039
S. limit of the NE. Trade .	5° 45′	5° 47′	11° 20′	9° 55′
N. limit of the SE. Trade .	2° 25′	1° 45′	3° 15′	3° 15′

Were the district of maximum rarefaction to reach
higher latitudes in summer, the SE. Trade would extend
still further over the equator into the northern hemisphere,
and its direction would be altered through south finally
into south-west, in consequence of its now passing over a
series of points, whose velocity of rotation is continually
decreasing. Under these circumstances, all places situated
in the northern portion of the torrid zone would lie in the
region of this southern Trade-wind, whose direction would
have been changed to south-west, as long as the sun
was on the northern side of the equator. During the
winter they would lie in the northern Trade-wind, and
the region of Calms would pass over them twice a year.
Thus their permanent Trade-wind would be changed into
a wind which changed periodically (a Monsoon). This
really happens in the Indian Ocean, and we must therefore submit the laws of its winds to a special investigation.

II.—The Yearly Periodical Winds.

1. THE MONSOONS OF THE EAST INDIES.

It would be more correct to describe the Trade-wind as an imperfectly developed monsoon, than to consider the latter as a modification of the former. *Monsun* (Malay *musim*) is derived from *mausim*, the Arabic for season. The Greeks learnt the fact of its existence during the expeditions under Alexander. Arrian says that according to the observations of Hippalus, who gave it the name of *Libonotus*, this wind appears in the Indian Ocean at the same time as the Etesian winds prevail in the Mediterranean, and that navigation from the ports was not practicable until the appearance of this southerly wind, which blew from the sea towards the land. In this he agrees with Aristotle, who describes expressly the regular alternation of the winds in those seas. Marco Polo first heard of it at Mangi, as the inhabitants of that place sail in winter to the Spice Islands near *Zipangri* (Ceylon), and return in summer with a wind in the opposite direction. The knowledge which the Arabs had of the phenomenon went very much into detail; for Sidi Ali, in his work Mohit, on the navigation of the Indian Ocean, compiled out of ten Arabic works, and published in 1554, gives the time of commencement of the Monsoon at fifty distinct places.

Halley (*Phil. Trans.* 1686, p. 158), describes them in the following words : —

Between the latitudes of 10° and 30° south, between Madagascar and Hollandia Nova, the general Trade-wind about the SE. by E. is found to blow all the year long, to all intents and purposes, after the same manner as in the same latitudes in the Ethiopic Ocean, as it is described in the fourth remark aforegoing.

The aforesaid SE. winds extend to within 2° of the equator, during the months of June, July, August, &c., to November, at which time between the south latitudes of 3° and 10°, being near the meridian of the north end of Madagascar, and between 2° and 12° south latitude, being near Sumatra and Java, the contrary winds from the NW. or between the north and west, set in and blow for half the year, viz., from the beginning of December till May; and this monsoon is observed as far as the Molucca Isles, of which more anon.

To the northward of 3° south latitude, over the whole Arabian or Indian Sea and Gulf of Bengal, from Sumatra to the coast of Africa, there is another monsoon, blowing from October to April upon the north-east points; but in the other half year, from April to October, upon the opposite points of SW. and WSW., and that with rather more force than the other, accompanied with dark rainy weather, whereas the NE. blows clear. 'Tis likewise to be noted that the winds are not so constant, either in strength or point, in the Gulf of Bengal, as they are in the Indian Sea, where a certain steady gale scarce ever fails. 'Tis also remarkable that the SW. winds in these seas are generally more southerly on the African side, more westerly on the Indian.

   .   .   .   .

To the eastward of Sumatra and Malacca, to the northward of the line, and along the coast of Camboia and China, the monsoons blow north and south,—that is to say, the NE. winds are much northerly, and the SW. much southerly. This constitution reaches to the eastwards of the Philippine Isles, and as far northerly as Japan. The northern monsoon setting in in these seas in October or November, and the southern in May, blowing all the summer months. Here it is to be noted that the points of the compass, from whence the wind comes in these parts of the world, are not so fixed as in those lately described; for the southerly will frequently pass a point or two to the eastwards of the south, and the northerly as much to the westwards of the north, which seems occasioned by the great quantity of land which is interspersed in these seas.

In the same meridians, but to the southwards of the equator, being that tract lying between Sumatra and Java to the west, and New Guinea to the east, the same northerly and southerly monsoons are observed,—but with this difference, that the inclination of the northerly is towards the NW., and of the southerly towards the SE.; but the *plagæ venti* are not more constant here than in the former, viz. variable 5 or 6 points. Besides, the times of the change of these winds are not the same as in the Chinese seas, but about a month or six weeks later.

These contrary winds do not shift all at once, but in some places the time of the change is attended with calms, in others with variable winds; and it is particularly remarkable that the end of the westerly monsoon on the coast of Coromandel, and the two last months of the southerly monsoon in the seas of China, are very subject to be tempestuous. The violence of these storms is such that they seem to be of the nature of the West India hurricanes, and render the navigation of these parts very unsafe about that time of the year. These tempests are by our seamen usually termed *the breaking-up of the monsoons.*

Capper (*Observations on the Winds and Monsoons*, London, 1801, p. 42) gives a more accurate account of these intermediate periods : —

On the Coromandel coast, between the end of one monsoon and the commencement of the other, the winds are variable, partaking of both directions. Calms often last through the whole of September, and into October. As soon as the sun commences to approach the zenith from the southward, the NE. monsoon loses its power, and there is a daily alternation of land and sea breezes,—a phenomenon not observed at its commencement. At this time the wind on the coast seems to follow the sun regularly, as it shifts through the whole compass every twenty-four hours. The violent storms do not occur at the breaking-up of the monsoon, but some time afterwards.

According to Horsburgh (*India Directory*, vol. ii. p. 200), the months of October and May are the months of change for the SW. and NE. monsoon north of the equator, and the months of April and October for the NW. and SE. monsoon south of the equator. He places the northern limit of the latter at lat. 2° N., and its southern limit at lat. 12° S. He places the extreme eastern limit of the district at 145° E. longitude near the Ladrone Islands. According to Goldingham's observations in Madras, carried on for a space of twenty-one years, the NE. monsoon lasts there from the 19th of October to the 2nd of March. Its setting-in varies from the 29th of September to the beginning of November. At Angara-kandy, on the Malabar coast, the SW. monsoon begins, according to Brown, on an average on the 31st of May, but varies between the 20th of May and 18th of June. According to Jansen, the change takes place in the following way in the Java seas : —

During the month of February the W. monsoon is still strong and steady ; in March it is interrupted by calms and squalls, which become less frequent and less violent in April. Now the easterly winds break in suddenly, clouds collect and darken the sky, while there are incessant thunder-storms by day and night, and waterspouts are very common. If the wind change again to W. or N. the sky clears again ; but this wind does not last, and the clouds soon reappear. The rain gradually ceases during the daytime, and SE. winds prevail throughout the month of May. At the time of the reverse change of the E. with the W. monsoon, the calms last for a shorter period, as the wind assumes a decided NW. direction at once, and the showers of rain, accompanied by violent squalls, are only felt for a short time. Thunder-storms are abundant, but only on land or close to the coast. Towards the end of November the NW. monsoon is permanent.

Observations carried on for a series of years (1850—1856) at Palembang, on the north coast of the south-eastern part of Sumatra, have led to the following results:—From November to March the prevalent winds are westerly and north-westerly. This is the regular rainy season, during the west monsoon. April is the month of change of the monsoon (*Kentering der Moesons*), when thunderstorms are most frequent. From May till September easterly and south-easterly winds (*Oostmoeson*) are predominant, and the change comes in September or October. From this it appears that the wind shifts pretty regularly round the compass, for its mean direction for each month, in order, counting from south to west, is 7° W., 20° W., 30° W., 28° N., 79° N., 85° N., 6° E., 21° E., 18° E., 25° E., 30° S., 4° W.

At Padang, the regular monsoon is nearly entirely concealed by the daily alternation of the land and sea-breezes, setting in nearly at right angles to the coast, which runs from NNW. to SSE. The land-wind is ENE., the sea-breeze WSW. The greater abundance of rain and the increased frequency of thunder-storms in the months of March and April, and in those of October and December, remind one of the change of the monsoons.

At Banjermassing, on the south coast of Borneo, the SW. monsoon prevails from December to March, the SE. monsoon from April to October. The change seems to be of short duration. Rain is most abundant from July to October, while thunder-storms are more frequent in the months of November, December, and May, at times consequently later than the changes of the monsoons. There is, however, in this respect, a considerable variation between individual years. In 1851, eighteen thunder-

storms were observed, while eighty-three took place in
1857. A closer examination of the direction of the wind
leads to the following results :—The predominant direction
of the wind in December is SW. and WSW., and it becomes more westerly in January and February. In March
the direction during the day is less constant. In April
the SE. wind becomes prevalent and increases in steadiness up to August and September. In October it gets
round to the southward. In November this is the case,
in the morning hours, in a still higher degree ; in fact, in
the afternoon, the wind goes somewhat past south towards
the west. At last in December the SW. monsoon is defi-
nitely established.

I take these determinations from Krecke's compendious
investigation of the observations contained in the ' Meteorological Observations in the Netherlands and their terri-
tories.' *

If we determine the mean directions of the wind, from
the ship's observations of individual directions collected
by Maury, by Lambert's formula, as Coffin has done for
the Atlantic Ocean, we obtain the following results : —

	North Latitude					South Latitude	
	25°—20°	20°—15°	15°—10°	10°—5°	5°—0°	0°—5°	5°—10°
January .	NE	NE	NE	NE	NNE	W	S
February	WNW	WSW	NE	NE	NNE	WNW	SW
March .	WSW	NNE	NE	NE	NE	WNW	SW
April . .	SSW	SSW	S	SSE	SW	WNW	SW
May . .	SSW	SSW	SSW	SW	SW	SW	SE
June . .	SSW	SW	SW	SW	SW	SSE	SE
July . .	SSW	SW	WSW	SW	SSW	SSW	SE
August .	SSW	SW	SW	SW	SW	S	SE
September	SSW	SSW	SW	SW	SW	SE	ESE
October .	ESE	NE	SSE	SW	WSW	S	ESE
November	NNW	NNE	NE	NE	WSW	WSW	NNE
December	NNE	NNE	NE	ENE	NNW	W	W

* *Meteorologische Waarnemingen in Nederland en zijne Bezittingen.*

The following table is more accurate : in it the angles of the direction of the wind are taken from S. through E. round to W. : —

	North Latitude					South Latitude	
	25°—20°	20°—15°	15°—10°	10°— 5°	5°— 0°	0°— 5°	5°—10°
January .	214° 26′	215° 53′	230° 51′	227° 13′	210° 14′	89° 55′	349° 49′
February .	119 1	68 14	227 29	226 3	209 47	108 7	53 45
March .	62 14	212 42	219 43	223 24	217 2	104 27	40 38
April . .	33 10	34 39	356 28	330 59	52 15	109 19	287 22
May . .	25 3	30 43	24 18	55 35	37 45	35 13	318 1
June . .	30 53	39 31	41 16	45 40	56 24	343 58	304 18
July . .	23 30	47 17	50 41	54 6	29 38	32 13	316 49
August .	30 15	42 40	52 0	35 10	47 46	359 49	306 50
September	23 24	48 10	52 15	49 43	42 3	306 0	299 35
October .	283 6	215 2	325 45	54 18	64 8	358 53	289 33
November	161 20	203 55	225 11	235 56	76 17	67 27	20 12
December	203 24	212 41	233 12	247 4	149 30	79 57	88 38

The next table gives the intensity, 100 being taken for the intensity if the wind had blown from the same quarter without interruption : —

	North Latitude					South Latitude	
	25°—20°	20°—15°	15°—10°	10°— 5°	5°— 0°	0°— 5°	5°—10°
January .	32	59	69	63	44	27	29
February .	13	14	59	74	49	39	29
March . .	59	48	39	73	37	21	17
April . .	71	68	11	29	33	18	25
May . .	84	75	66	76	74	22	38
June . .	83	83	93	81	83	49	49
July . .	70	84	93	70	70	54	75
August .	87	74	82	65	67	48	80
September	28	57	78	91	65	41	58
October .	30	22	18	37	61	16	65
November	29	69	68	13	48	35	9
December	79	77	73	39	16	42	31
Mean . .	55·4	60·8	62·4	59·3	53·9	35·2	42·1

E

We see from this that the minimum intensity of motion is in the vicinity of the equator, and this fact is more clearly marked to the south than to the north of the line. We cannot here speak of a zone of Calms, as in the region of the Trade-winds, inasmuch as there is a constant current during the summer months, the direction of which changes, on crossing the line, from that of the SE. Trade-wind to one which is south-westerly. We see, however, that the line of partition between the northern and southern hemisphere falls in this district within the latter, if we assume the least value of the mean intensity of both currents to indicate that line.

The Dutch observations made at Nangasaki in Japan (1845—55) afford us accurate information about the winds at the outer limit of the Monsoon area. In the month of September the wind is NNE. and in October it gets round toward N. From November to January the mean direction varies a few degrees from N. on the west side, and becomes more westerly in the succeeding months, so that it is nearly WNW. in April. In June, however, it is nearly NNW. again. From June to August the prevalent winds are SW., and they are most southerly in July. The months of change of the Monsoons are accordingly April and May on the one hand, and August and September on the other. Thunder-storms are not common; they are most frequent in March, July, and August.

We have seen that the Trade-winds set in from both sides to that parallel of latitude within the torrid zone at which the total pressure of the atmosphere is least. This result accords perfectly with a hypothesis, which must necessarily be admitted to be allowable, that the air rises there. In former works I have demonstrated that the same fact is true of the Monsoons, and it will therefore suffice for my present purpose to quote from them the following statements:—

The dry air, and the aqueous vapour disseminated through it, exert a pressure in common on the barometer. Hence we see that the barometrical column consists of two portions, of which one is due to the dry air, the other to the aqueous vapour. From this it is clear that, since on a rise of temperature, the air increases in volume, rises, and flows away in the upper strata of the atmosphere, while, owing to the same rise of temperature, the evaporation is increased, and with it, the elastic force of the vapour contained in the air; we shall be able to trace an evident connection between the periodical variations of the barometer and of temperature. Until we can ascertain the quantitative relation of both these variations, which take place simultaneously, but in opposite directions, we are unable to determine whether the total pressure is increased or diminished by a rise in temperature, or whether at one period the total variation is not in the one direction, and at another in that which is opposite. We must therefore consider the variations of the pressure of dry air and of aqueous vapour separately, in order to be able to understand the periodical barometrical observations. For this we have the following rules : —

(1.) At all points of observation, in the torrid and temperate zones, the elastic force of the aqueous vapour contained in the air increases with a rise of temperature. This increase from the colder to the warmer months is greatest in the region of the Monsoons, especially towards the northern limits of that area. The form of the curve, which represents the elastic force in this area, is not decidedly convex, for the tension of aqueous vapour undergoes such a slight alteration during the SW. Monsoon, that it remains nearly constant for several months, and accordingly the vertex of the curve becomes a right line. In the vicinity of the equator, the change from the convex curve of the northern hemisphere to the concave curve of the southern is observable at Buitenzorg in Java. The point where this change takes place in the Atlantic Ocean appears to lie farther to the north of the equator.

The following table* gives the tension of aqueous vapour in English inches for a series of stations situated either actually in the region of the northern Monsoon, or in its immediate neighbourhood. The results have been obtained by hygrometrical observations. The pressure of

TENSION OF AQUEOUS VAPOUR.

	Jan.	Feb.	Mar.	April	May	June	July	Aug.	Sept.	Oct.	Nov.	Dec.	Year.
Yakutsk . .	...	...	...	...	0·142	0·294	0·364	0·323	0·228	0·083	...	...	...
Baganida, 1 .	...	...	...	0·041	0·085	0·179	0·312	0·274	0·149	0·098	...	...	...
Ajansk, 2 . .	0·037	0·060	0·082	0·112	0·175	0·245	0·343	0·366	0·279	0·134	0·074	0·045	0·162
Pekin, 12 . .	0·080	0·099	0·134	0·216	0·324	0·506	0·714	0·664	0·447	0·251	0·138	0·090	0·306
Nangasaki. .	0·234	0·241	0·222	0·402	0·479	0·615	0·795	0·827	0·709	0·487	0·346	0·272	0·470
Hongkong, 1 .	0·371	0·357	0·515	0·698	0·817	0·906	0·917	0·900	0·842	0·705	0·660	0·447	0·678
Nertschinsk, 9	0·014	0·023	0·057	0·107	0·175	0·344	0·444	0·279	0·235	0·095	0·045	0·022	0·153
Barnaoul, 9 .	0·048	0·058	0·094	0·153	0·213	0·336	0·433	0·381	0·256	0·164	0·087	0·064	0·191
Tomsk, 1½ . .	0·063	0·043	0·078	0·153	0·243	0·367	0·476	0·458	0·296	0·151	0·061	0·075	0·206
Bogoslowsk .	0·045	0·057	0·078	0·120	0·180	0·283	0·395	0·321	0·242	0·149	0·088	0·053	0·168
Slatoust . .	0·053	0·073	0·090	0·150	0·220	0·328	0·416	0·343	0·251	0·160	0·101	0·066	0·188
Catherinenburg	0·057	0·071	0·089	0·137	0·209	0·314	0·410	0·344	0·252	0·163	0·104	0·072	0·185
Nishne Tagilsk	0·050	0·080	0 101	0·166	0·219	0·340	0·488	0·421	0·260	0·193	...	...	...
Orenburg, 9 .	0·074	0·068	0·092	0·158	0·282	0·344	0·406	0·375	0·257	0·161	0·100	0·082	0·200
Derbent, 4½ .	0·178	0·193	0·216	0·279	0·429	0·542	0·634	0·642	0·497	0·428	0·345	0·223	0·384
Baku, 7¾ . .	0·201	0·201	0·220	0·299	0·551	0·680	0·701	0·547	0·547	0·451	0·313	0·243	0·413
Lenkoran, 7 .	0·209	0·236	0·272	0·357	0·517	0·615	0·701	0·698	0·581	0·476	0·340	0·229	0·436
Alexandropol.	0·085	0·098	0·124	0·182	0·272	0·351	0·367	0·362	0·281	0·225	0·160	0·126	0·220
Alagir, 2 . .	0·148	0·131	0·168	0·234	0·373	0·441	0·506	0·491	0·362	0·310	0·193	0·166	0·293
Tiflis, 13 . .	0·149	0·161	0·184	0·251	0·363	0·419	0·476	0·478	0·400	0·323	0·239	0·172	0·301
Kutais, 3 . .	0·190	0·206	0·225	0·305	0·454	0·551	0·632	0·678	0·539	0·407	0·288	0·207	0·389
Redout Kale, 8	0·194	0·206	0·237	0·322	0·436	0·558	0·647	0·696	0·537	0·431	0·302	0·218	0·398
Raimsk . .	0·089	0·046	0·142	0·228	0·252	0·306	0·366	0·350	0·303	0·216	0·126	0·082	0·209
Calcutta, 4 .	0·582	0·576	0·740	0·839	0·925	0·958	0·969	0·961	0·952	0·851	0·634	0·498	0·791
Bombay, 4 .	0·588	0·630	0·728	0·821	0·873	0·940	0·920	0·912	0·855	0·877	0·740	0·653	0·794
Madras, 5 . .	0·698	0·707	0·811	0·914	0·898	0·849	0·813	0·837	0·866	0·858	0·731	0·721	0·809
Trevandrum, 9	0·723	0·739	0·802	0·862	0·862	0·839	0·814	0·801	0·801	0·823	0·805	0·758	0·802
Amboyna . .	0·869	0·863	0·859	0·858	0·859	0·853	0·813	0·822	0·826	0·822	0·844	0·832	0·844
Palembang .	0·865	0·854	0·865	0·891	0·880	0·879	0·834	0·819	0·817	0·826	0·849	0·853	0·853
Buitenzorg .	0·756	0·756	0·757	0·755	0·761	0·741	0·717	0·713	0·713	0·731	0·746	0·739	0·741
Banjoewangie	0·916	0·912	0·940	0·921	0·913	0·897	0·854	0·847	0·872	0·898	0·895	0·896	0·897

* *Vide* also The Sixth Number of Meteorological Papers, published by authority of the Board of Trade, 1861.

the dry air has been obtained from the height of the barometer by deducting therefrom the tension of the aqueous
vapour. The numbers which follow some of the names
indicate the number of years from which the mean values
have been taken.

(2.) The variation in the pressure of the dry air takes
place in the opposite direction to that of the aqueous
vapour, as it diminishes from the colder to the warmer
months. The maximum diminution always occurs in the
warmest month in the temperate zone, and consequently
it is observed in the northern hemisphere in July, in the
southern in January or February. The oscillation is of
greatest magnitude at the northern limit of the northern
Monsoon, reaching a maximum of 1·387 in. in Pekin. It
exceeds an inch at Hong Kong, Benares, and Barnaoul,
nearly reaches that value at Calcutta and Yakutsk, and
is still above 0·9 in. at the Caspian Sea. In Australia it
does not exceed 0·75 of an inch, and in Western Europe
it is about 0·3.

In the following table the pressure of dry air in each
month is given as the variation from the mean pressure
for the year :—

PRESSURE OF DRY AIR.

(1) NORTHERN ASIA.

	Jan.	Feb.	March	April	May	June	July	August	Sept.	Oct	Nov.	Dec.	Mean	Oscillation
Yakutsk	+0·333	+0·395	+0·169	−0·026	−0·223	−0·480	−0·535	−0·352	−0·070	+0·035	+0·257	+0·498	29·553	1·033
Ajansk	+0·132	+0·144	+0·209	+0·076	−0·033	−0·175	−0·328	−0·232	−0·131	+0·102	+0·149	+0·087	29·708	0·538
Pekin	+0·570	+0·580	+0·272	+0·026	−0·233	−0·565	−0·818	−0·661	−0·217	+0·169	+0·542	+0·323	29·728	1·398
Nangasaki	+0·425	+0·371	+0·365	+0·067	−0·106	−0·344	−0·617	−0·592	−0·352	+0·032	+0·283	+0·386	29·512	1·017
Hongkong	+0·537	+0·529	+0·267	−0·019	−0·192	−0·388	−0·561	−0·436	−0·293	+0·037	+0·097	+0·434	29·249	1·098
Nertschinsk (a)	+0·357	+0·318	+0·197	+0·018	−0·198	−0·374	−0·501	−0·348	−0·095	+0·126	+0·234	+0·261	27·620	0·858
" (b)	+0·357	+0·286	+0·177	−0·006	−0·183	−0·351	−0·478	−0·343	−0·047	+0·110	+0·210	+0·264	27·639	0·835
Barnaoul (a)	+0·445	+0·366	+0·224	+0·079	−0·142	−0·498	−0·596	−0·457	−0·137	+0·111	+0·296	+0·341	29·394	1·041
" (b)	+0·411	+0·358	+0·209	+0·058	−0·048	−0·422	−0·610	−0·467	−0·120	+0·107	+0·306	+0·317	29·399	1·021
Tomsk	+0·462	+0·605	+0·288	+0·093	−0·184	−0·428	−0·733	−0·500	−0·184	+0·086	+0·312	+0·185	29·656	1·338
Bogoslowsk	+0·197	+0·132	+0·134	+0·047	−0·017	−0·264	−0·351	−0·185	−0·064	+0·050	+0·168	+0·163	29·101	0·548
Slatoust	+0·223	+0·154	+0·136	+0·062	−0·059	−0·295	−0·398	−0·223	−0·029	+0·083	+0·199	+0·147	28·415	0·621
Catherinenburg (a)	+0·224	+0·172	+0·121	+0·063	−0·065	−0·368	−0·464	−0·312	−0·018	+0·121	+0·211	+0·265	28·839	0·729
" (b)	+0·238	+0·120	+0·105	+0·017	0·000	−0·250	−0·374	−0·209	−0·003	+0·056	+0·209	+0·092	28·847	0·612
Nishne Tagilsk	+0·293	+0·178	+0·158	+0·040	−0·042	−0·307	−0·449	−0·292	−0·009	+0·054	+0·221	+0·161	29·082	0·742
F. Alexandrowsk	+0·445	+0·402	+0·216	+0·051	−0·158	−0·304	−0·554	−0·539	−0·090	+0·253	+0·145	+0·133	29·745	0·999
Orenburg	+0·319	+0·201	+0·185	+0·068	−0·137	−0·356	−0·501	−0·336	−0·077	+0·116	+0·310	+0·207	29·616	0·820
Lugan	+0·337	+0·111	+0·087	−0·028	−0·109	−0·210	−0·441	−0·256	−0·047	+0·108	+0·260	+0·242	29·469	0·778

(2) CAUCASUS AND TARTARY.

	Jan.	Feb.	March	April	May	June	July	August	Sept.	Oct.	Nov.	Dec.	Mean	Oscillation
Derbent	+0·351	+0·204	+0·188	+0·090	-0·082	-0·336	-0·460	-0·364	-0·132	+0·080	+0·181	+0·278	29·741	0·811
Baku	+0·335	+0·272	+0·228	+0·076	-0·102	-0·309	-0·497	-0·453	-0·161	+0·091	+0·250	+0·272	29·772	0·832
Lenkoran	+0·391	+0·244	+0·186	+0·031	-0·144	-0·348	-0·492	-0·428	-0·177	+0·079	+0·337	+0·322	29·781	0·883
Novo Petrovsk	+0·299	+0·263	+0·182	+0·025	-0·166	-0·349	-0·505	-0·380	-0·121	+0·063	+0·317	+0·261	29·729	0·822
Aralsk	+0·454	+0·250	+0·254	-0·021	-0·170	-0·527	-0·654	-0·383	-0·230	+0·225	+0·488	+0·299	29·307	1·142
Aralich	+0·386	+0·212	+0·057	-0·074	-0·074	-0·277	-0·334	-0·349	-0·119	+0·109	+0·207	+0·265	26·992	0·735
Raimsk, 3	+0·366	+0·308	+0·219	+0·009	-0·143	-0·408	-0·681	-0·392	-0·105	+0·072	+0·333	+0·298	29·358	1·047
Alexandropol, 6½	+0·149	+0·045	+0·023	-0·035	-0·066	-0·180	-0·190	-0·141	-0·008	+0·160	+0·157	+0·092	24·779	0·350
Alagir	+0·141	-0·108	+0·099	+0·007	-0·120	-0·214	-0·303	-0·248	-0·019	+0·145	+0·236	+0·164	27·685	0·539
Tiflis	+0·255	+0·129	+0·113	-0·001	-0·121	-0·234	-0·319	-0·267	-0·081	+0·093	+0·215	+0·220	29·128	0·574
Kutais, 3	+0·340	+0·217	+0·169	+0·040	-0·090	-0·258	-0·398	-0·407	-0·150	+0·053	+0·190	+0·296	29·183	0·747
Redout Kale, 8	+0·288	+0·156	+0·149	-0·023	-0·109	-0·303	-0·442	-0·478	-0·209	-0·002	+0·143	+0·239	29·699	0·766
Petigorsk, 2	+0·175	+0·185	+0·110	+0·012	-0·170	-0·234	-0·327	-0·234	-0·018	+0·029	+0·129	+0·170	27·996	0·512
Trebizond	+0·169	+0·189	+0·176	+0·061	-0·019	-0·350	-0·215	-0·215	-0·175	+0·065	+0·064	+0·248	29·577	0·598
Constantinople	+0·234	+0·097	+0·105	-0·034	-0·063	-0·230	-0·244	-0·264	-0·069	+0·081	+0·209	+0·179	29·497	0·498

(3) HINDOSTAN.

	Jan.	Feb.	March	April	May	June	July	August	Sept.	Oct.	Nov.	Dec.	Mean	Oscillation
Calcutta, 4 (a)	+0·457	+0·406	+0·145	-0·070	-0·258	-0·390	-0·419	-0·346	-0·246	+0·002	+0·359	+0·557	28·977	0·978
" (b)	+0·520	+0·372	+0·112	-0·124	-0·265	-0·434	-0·427	-0·375	-0·247	+0·016	+0·301	+0·546		0·980
Benares	+0·528	+0·440	+0·343	+0·162	-0·080	-0·561	-0·701	-0·660	-0·363	-0·061	+0·309	+0·642	28·797	1·243
Nusseerabad	+0·438	+0·326	+0·256	+0·145	+0·005	-0·427	-0·626	-0·571	-0·345	+0·130	+0·304	+0·365	27·838	1·064
Madras	+0·268	+0·230	+0·047	-0·119	-0·221	-0·195	-0·135	-0·125	-0·116	-0·027	+0·179	+0·209	29·488	0·489
Trevandrum	+0·128	+0·094	-0·002	-0·096	-0·119	+0·112	-0·023	-0·011	+0·009	-0·003	+0·015	+0·084	28·887	0·247
Bombay, 4	+0·330	+0·270	+0·119	-0·046	-0·145	-0·315	-0·288	-0·276	-0·090	-0·071	+0·104	+0·292	29·048	0·645
Poonah	+0·293	+0·285	+0·231	+0·029	-0·244	-0·390	-0·366	-0·280	-0·166	-0·049	+0·184	+0·370	27·526	0·760
Mercara	+0·165	+0·119	-0·012	-0·034	-0·068	-0·128	-0·208	-0·114	-0·117	-0·074	+0·078	+0·289	25·557	0·497
Dodabetta	+0·098	+0·086	+0·112	+0·019	-0·063	-0·098	-0·096	-0·074	-0·062	-0·116	+0·038	+0·065	21·634	0·228
Colombo	+0·128	+0·096	+0·010	-0·072	-0·098	-0·080	-0·002	-0·043	-0·031	-0·018	-0·030	+0·138	29·006	0·237

(4) SOUTHERN HEMISPHERE.

	Jan.	Feb.	March	April	May	June	July	August	Sept.	Oct.	Nov.	Dec.	Mean	Oscillation
Buitenzorg	−0·005	−0·007	−0·018	−0·021	−0·033	−0·003	+0·018	+0·031	+0·035	+0·026	+0·022	−0·002	28·299	0·581
Souillac	−0·296	−0·247	−0·112	−0·041	+0·083	+0·193	+0·285	+0·279	+0·142	+0·050	−0·149	−0·186	29·112	0·713
Port Jackson . .	−0·339	−0·252	−0·049	−0·031	+0·137	+0·268	+0·374	+0·258	+0·052	−0·077	−0·131	−0·211	28·938	0·378
Melbourne . . .	−0·134	−0·148	−0·030	+0·109	+0·167	+0·137	+0·024	+0·196	−0·040	−0·100	−0·182	−0·026	29·384	0·378
Hobarton . . .	−0·086	−0·033	+0·009	+0·029	+0·052	+0·080	+0·088	+0·048	+0·009	+0·001	−0·129	−0·074	29·479	0·217
Cape Town . . .	−0·234	−0·240	−0·144	−0·042	+0·102	+0·196	+0·236	+0·216	+0·147	+0·036	−0·088	−0·187	29·684	0·476
Rio Janeiro . . .	−0·192	−0·217	−0·159	−0·105	+0·117	+0·229	+0·341	+0·173	+0·131	+0·023	−0·095	−0·154	29·141	0·558
St. Jago . . .	−0·100	−0·121	−0·070	−0·010	+0·026	+0·091	+0·083	+0·112	+0·067	+0·042	−0·037	−0·079	28·411	0·233

3. The periodical variations of atmospherical pressure follow immediately from the combined action of these two fluctuations. Throughout the whole of Asia the curve which represents the barometrical pressure agrees very closely with that which represents the pressure of dry air during the entire year; i.e., it has a concave form and reaches a minimum in July. The tendency thereto is perceptible in Russia in Europe, as far west as the meridian of St. Petersburg, and becomes more clearly marked as we approach the Ural Mountains. This will be seen from the following table :—

FLUCTUATIONS OF THE BAROMETER.

(1) NORTHERN ASIA.

	Jan.	Feb.	March	April	May	June	July	August	Sept.	Oct.	Nov.	Dec.	Mean	Oscillation
Yakutsk, 1⅔	+0·216	+0·278	+0·069	−0·068	−0·207	−0·312	−0·296	−0·244	+0·032	−0·009	+0·149	+0·381	29·678	0·693
Udskoi, 1	+0·312	+0·175	+0·074	+0·074	−0·100	−0·140	−0·228	−0·278	−0·122	−0·036	+0·097	+0·169	29·689	0·590
Ajansk, 2	+0·007	+0·042	+0·129	+0·194	−0·029	−0·092	−0·148	−0·027	−0·015	+0·073	+0·061	−0·030	29·871	0·342
Pekin, 12	+0·344	+0·274	+0·100	−0·063	−0·214	−0·366	−0·408	−0·294	−0·074	+0·114	+0·256	+0·327	30·154	0·752
Nangasaki	+0·189	+0·141	+0·116	−0·001	−0·096	−0·200	−0·193	−0·235	−0·113	+0·049	+0·156	+0·187	29·983	0·424
Hakodati	+0·036	+0·181	+0·159	+0·082	−0·013	−0·155	−0·119	−0·129	−0·068	+0·081	+0·091	−0·155	29·887	0·336
Nafa, 1	+0·151	+0·145	+0·127	+0·055	+0·061	−0·133	−0·155	−0·258	−0·153	−0·018	+0·128	+0·179	29·935	0·437
Macao	+0·186	+0·151	+0·114	−0·049	−0·060	−0·231	−0·207	−0·207	−0·037	+0·008	+0·118	+0·216	30 047	0·447
Canton	+0·280	+0·204	+0·123	−0·045	−0·134	−0·164	−0·239	−0·236	−0·210	+0·016	+0·177	+0·228	29·594	0·519
Hongkong	+0·230	+0·208	+0·104	−0·002	−0·063	−0·160	−0·323	−0·215	−0·029	+0·064	+0·079	+0·203	29·927	0·563
Shanghae	+0·331	+0·292	+0·135	−0·058	−0·202	−0·347	−0·403	−0·372	−0·081	+0·111	+0·264	+0·326	30·111	0·734
Nertschinsk, 17	+0·208	+0·177	+0·091	−0·038	−0·185	−0·193	−0·221	−0·132	−0·023	+0·076	+0·116	+0·120	27·781	0·429
9	+0·208	+0·145	+0·071	−0·062	−0·171	−0·169	−0·197	−0·128	+0·025	+0·060	+0·093	+0·122	27·799	0·404
Barnaoul, 20	+0·302	+0·232	+0·103	+0·041	−0·120	−0·352	−0·355	−0·266	−0·071	+0·083	+0·192	+0·214	29·585	0·657
9	+0·269	+0·224	+0·112	+0·020	−0·026	−0·278	−0·367	−0·276	−0·064	+0·080	+0·201	+0·189	29·590	0·636
Irkutsk, 15	+0·275	+0·216	+0·121	−0·045	−0·151	−0·244	−0·312	−0·240	−0·061	+0·085	+0·153	+0·175	29·273	0·587
Tomsk	+0·319	+0·442	+0·161	+0·041	−0·146	−0·265	−0·463	−0·249	−0·090	+0·031	+0·168	+0·056	29·862	0·905
Tobolsk, 6	+0·247	+0·255	+0·137	+0·018	+0·009	−0·198	−0·231	−0·143	−0·166	−0·036	+0·085	+0·112	29·839	0·485
Bogoslowsk, 17	+0·074	+0·021	+0·044	−0·001	−0·005	−0·149	−0·124	−0·033	+0·009	+0·032	+0·089	+0·048	29·270	0·238
Sladoust, 19	+0·097	+0·039	+0·039	+0·014	−0·026	−0·154	−0·169	−0·066	+0·034	+0·056	+0·112	+0·026	29·101	0·281
Catharinenburg, 22	+0·097	+0·032	+0·016	+0·007	−0·054	−0·188	−0·183	−0·096	+0·090	+0·070	+0·093	+0·117	29·025	0·305
9	+0·110	+0·006	+0·009	−0·031	+0·024	−0·121	−0·149	−0·049	+0·063	+0·034	+0·129	−0·021	29·034	0·278
Nishne Tagilsk	+0·135	+0·050	+0·051	−0·002	−0·031	−0·175	−0·168	−0·078	+0·042	+0·039	+0·117	+0·015	29·290	0·310
Berezow	+0·062	+0·071	+0·206	+0·165	−0·163	−0·153	−0·150	−0·172	−0·091	−0·024	+0·063	−0·014	29·768	0·369
Tscherdyn, 1	+0·219	−0·151	+0·269	−0·096	−0·137	−0·225	−0·202	+0·029	+0·264	−0·333	+0·198	+0·438	29·415	0·459
Orenburg, 13	+0·173	+0·070	+0·034	+0·029	−0·061	−0·223	−0·276	−0·145	−0·019	+0·078	+0·183	+0·106	29·799	0·475
10	+0·195	+0·160	+0·079	+0·028	−0·053	−0·209	−0·263	−0·158	−0·018	+0·079	+0·212	+0·092	29·813	
F. Alexandrowsk	+0·217	+0·192	+0·050	−0·013	−0·117	−0·156	−0·245	−0·152	+0·010	+0·113	+0·071	−0·050	30·047	0·462

(2) CAUCASUS AND TARTARY.

	Jan.	Feb.	March	April	May	June	July	August	Sept.	Oct.	Nov.	Dec.	Mean	Oscillation
Derbent	+0·153	+0·016	+0·023	−0·010	−0·036	−0·174	−0·196	−0·112	−0·016	+0·128	+0·101	+0·120	30·123	0·349
Baku	+0·132	+0·069	+0·042	−0·029	−0·066	−0·161	−0·221	−0·154	−0·018	+0·137	+0·159	+0·112	30·177	0·380
Lenkoran	+0·172	+0·052	+0·031	−0·039	−0·055	−0·161	−0·217	−0·158	−0·024	+0·127	+0·149	+0·123	30·210	0·389
Novo Petrowsk, 6	+0·125	+0·066	+0·035	−0·046	−0·082	−0·175	−0·223	−0·151	−0·023	+0·161	+0·189	+0·124	30·036	0·412
Aralsk, 2	+0·247	+0·060	+0·108	−0·056	−0·089	−0·357	−0·439	−0·133	−0·056	+0·260	+0·356	+0·096	29·582	0·795
Aralich, 2¾	+0·225	+0·062	−0·033	−0·066	−0·031	−0·151	−0·209	−0·166	−0·018	+0·107	+0·144	+0·132	27·313	0·434
Raimsk	+0·248	+0·148	+0·154	+0·032	−0·097	−0·309	−0·420	−0·249	−0·009	+0·182	+0·251	+0·175	29·562	0·671
Alexandropol	+0·015	−0·074	−0·074	−0·079	−0·016	−0·040	−0·045	+0·002	+0·053	+0·163	+0·103	−0·005	25·021	
Alagir	−0·003	−0·054	−0·026	−0·052	−0·041	−0·066	−0·090	−0·049	+0·049	+0·161	+0·137	+0·037	27·977	
Tiflis, 11	+0·102	−0·010	−0·003	−0·052	−0·061	−0·116	−0·146	−0·091	+0·017	+0·124	+0·151	+0·090	28·637	0·297
Kutais, 3	+0·140	+0·026	+0·006	−0·045	−0·225	−0·096	−0·166	−0·119	−0·001	+0·071	+0·089	+0·114	29·574	0·365
Redout Kale	+0·134	+0·013	+0·036	−0·050	−0·022	−0·093	−0·142	−0·132	−0·022	+0·080	+0·096	+0·108	30·049	0·276
Petigorsk	+0·030	−0·035	−0·042	−0·054	−0·080	−0·070	−0·143	−0·045	+0·058	+0·213	+0·048	+0·049	28·285	

(3) EASTERN EUROPE.

	Jan.	Feb.	March	April	May	June	July	August	Sept.	Oct.	Nov.	Dec.	Mean	Oscillation
Trebizond	+0·002	+0·027	+0·040	−0·003	+0·010	−0·154	−0·056	−0·056	−0·001	+0·140	+0·137	+0·105	29·867	
Constantinople	+0·066	+0·039	+0·071	−0·110	−0·043	−0·090	−0·098	−0·082	−0·008	+0·090	+0·161	+0·071	29·942	
Kasan, 13	+0·092	+0·083	+0·064	+0·045	−0·033	−0·138	−0·201	−0·124	−0·000	+0·086	+0·019	+0·106	29·713	0·307
Lugan, 14	+0·180	−0·014	−0·017	−0·059	−0·053	−0·169	−0·200	−0·100	+0·014	+0·103	+0·177	+0·117	29·631	0·380
Nicholaieff, 12	+0·120	+0·039	+0·002	+0·055	−0·073	−0·102	−0·126	−0·072	+0·027	+0·052	+0·082	+0·103	28·886	0·246
Odessa, 10	+0·088	+0·004	+0·023	−0·055	−0·049	−0·087	−0·103	−0·062	−0·002	+0·050	+0·126	+0·164	29·879	0·267
Kaluga, 6	+0·063	−0·024	+0·022	−0·040	−0·005	−0·066	−0·103	−0·041	−0·010	+0·131	+0·122	−0·066	29·468	0·234
Moscow, 14	+0·135	−0·082	−0·041	−0·031	−0·014	−0·178	−0·135	−0·003	+0·099	+0·050	+0·097	+0·104	29·620	0·313

(4) AFRICA.

	Jan.	Feb.	March	April	May	June	July	August	Sept.	Oct.	Nov.	Dec.	Mean	Oscillation
Aden	+0·118	+0·138	+0·072	−0·003	−0·095	−0·177	−0·222	−0·193	−0·072	+0·074	+0·171	+0·185	29·704	0·407
Cairo	+0·097	+0·132	−0·004	−0·082	−0·062	−0·079	−0·173	−0·149	+0·019	+0·035	+0·095	+0·176	29·903	0·349
Montaganem, 2	+0·095	+0·097	−0·035	−0·029	−0·094	−0·001	−0·096	−0·041	−0·016	−0·017	−0·003	+0·142	29·684	0·238
Oran, 12	+0·086	+0·058	+0·025	−0·040	−0·063	−0·029	−0·038	−0·063	−0·038	−0·042	+0·035	+0·102	29·899	0·165
Algiers	+0·024	+0·017	−0·002	−0·010	−0·011	−0·046	−0·025	−0·007	+0·006	+0·074	−0·016	+0·066		0·120

(5) HINDOSTAN.

	Jan.	Feb.	March	April	May	June	July	August	Sept.	Oct.	Nov.	Dec.	Mean	Oscillation
Bombay, 5	+0·126	+0·111	+0·058	−0·016	−0·063	−0·166	−0·169	−0·092	−0·024	+0·018	+0·080	+0·130	29·815	0·296
Poonah	+0·166	+0·080	+0·030	−0·014	−0·162	−0·172	−0·135	−0·080	−0·003	+0·027	+0·121	+0·145	27·921	0·337
Mercara	+0·057	+0·081	+0·032	+0·005	−0·009	−0·036	−0·044	−0·079	−0·043	−0·030	+0·024	+0·047	26·071	0·160
Seringapatam	+0·166	+0·115	+0·036	−0·044	−0·063	−0·092	−0·076	−0·076	−0·066	+0·014	+0·035	+0·048	27·456	0·257
Nusseerabad	+0·239	+0·161	+0·081	−0·011	−0·127	−0·239	−0·262	−0·211	−0·098	+0·069	+0·195	+0·215	28·235	0·500
Mozufferpoor	+0·236	+0·176	+0·061	−0·020	−0·168	−0·218	−0·248	−0·198	−0·122	+0·044	+0·222	+0·230	29·453	0·484
Nusseera	+0·208	+0·157	+0·060	−0·007	−0·089	−0·203	−0·209	−0·168	−0·129	+0·052	+0·148	+0·183	29·180	0·417
Benares	+0·272	+0·174	+0·106	−0·045	−0·137	−0·289	−0·308	−0·203	−0·098	+0·074	+0·180	+0·275	29·468	0·683
Calcutta, 10	+0·388	+0·209	+0·050	−0·051	−0·139	−0·253	−0·304	−0·230	−0·105	+0·006	+0·137	+0·285	29·713	0·692
4	+0·233	+0·175	+0·080	−0·038	−0·140	−0·240	−0·256	−0·191	−0·100	+0·048	+0·187	+0·248	29·783	0·504
Agra	+0·254	+0·197	+0·112	−0·104	−0·121	−0·293	−0·264	−0·210	−0·120	+0·091	+0·191	+0·272	29·233	0·565
Madras, 5	+0·168	+0·129	+0·049	−0·014	−0·132	−0·154	−0·130	−0·095	−0·058	+0·022	+0·102	+0·122	29·841	0·312
Trevandrum	+0·049	+0·030	−0·002	−0·037	−0·059	−0·040	−0·011	−0·013	+0·002	+0·017	+0·018	+0·040	29·690	0·108
Singapore	−0·033	+0·030	−0·000	+0·002	+0·012	−0·026	−0·016	−0·004	+0·002	+0·013	−0·018	−0·000	29·883	0·059
Manilla	+0·139	+0·084	+0·043	+0·002	−0·055	−0·089	−0·088	−0·102	−0·093	−0·033	+0·035	+0·163	29·851	0·255
Canton	+0·281	+0·204	+0·123	−0·045	−0·134	−0·164	−0·239	−0·236	−0·210	+0·016	+0·177	+0·228	29·894	0·520
Macao	+0·186	+0·151	+0·114	−0·048	−0·060	−0·231	−0·207	−0·207	−0·237	+0·008	+0·118	+0·216	30·046	0·446
Tirhoot	+0·307	+0·121	+0·026	−0·084	−0·137	−0·244	−0·265	−0·217	−0·153	+0·055	+0·180	+0·224	29·389	0·572
Ava	+0·228	+0·114	+0·061	−0·027	−0·105	−0·156	−0·177	−0·126	−0·100	+0·010	+0·201	+0·245	29·573	0·405
Serampore	+0·274	+0·219	+0·151	−0·061	−0·054	−0·217	−0·397	−0·278	−0·158	+0·047	+0·209	+0·245	28·765	0·671
Mahabuleshwur	+0·053	+0·081	+0·004	−0·017	−0·041	−0·019	−0·084	−0·037	−0·160	−0·126	+0·049	+0·055	25·082	0·210
Ootacamund	+0·230	+0·121	+0·074	+0·022	−0·051	−0·107	−0·151	−0·130	−0·081	−0·001	+0·029	+0·218	23·050	0·381
Dodabetta	+0·035	+0·052	+0·086	+0·051	+0·009	−0·080	−0·072	−0·042	−0·042	+0·004	−0·031	−0·002	22·045	0·166

(6) NORTH ATLANTIC OCEAN.

Latitude N	Jan.	Feb.	March	April	May	June	July	August	Sept.	Oct.	Nov.	Dec.	Mean	Oscillation
35°—30°	+0·032	+0·001	−0·147	−0·073	+0·029	+0·042	+0·033	−0·024	−0·059	−0·027	−0·174	−0·029		
30—25	+0·019	+0·082	−0·027	+0·016	+0·071	+0·073	+0·038	+0·012	−0·052	−0·012	−0·125	+0·027		
25—20	+0·015	+0·028	+0·024	+0·016	+0·058	+0·076	+0·010	−0·006	−0·053	−0·010	−0·068	−0·087		
20—15	+0·008	+0·045	+0·023	−0·016	+0·037	+0·016	−0·018	−0·018	−0·023	−0·023	−0·034	−0·010		
15—10	−0·013	+0·003	+0·021	+0·005	+0·029	−0·007	+0·008	−0·030	−0·033	−0·016	−0·005	−0·008		
10—5	+0·018	+0·012	−0·019	−0·008	+0·010	−0·007	+0·035	+0·009	+0·010	−0·008	−0·016	−0·014		
5—0	−0·032	0·000	−0·038	−0·027	−0·005	−0·017	+0·054	+0·028	+0·069	+0·027	−0·001	−0·010		
Mean	−0·001	+0·025	−0·025	−0·010	+0·033	+0·025	+0·024	−0·005	−0·021	−0·010	−0·056	−0·010		
St. Michael, 10 . .	+0·016	−0·035	+0·054	+0·041	+0·022	+0·028	+0·077	−0·002	+0·066	−0·078	−0·095	+0·000	30·122	
Funchal, 3 . . .	+0·042	−0·022	+0·027	−0·082	+0·010	+0·022	−0·004	+0·001	−0·010	+0·006	−0·034	+0·095	30·015	

(7) SOUTH ATLANTIC OCEAN.

Latitude S	Jan.	Feb.	March	April	May	June	July	August	Sept.	Oct.	Nov.	Dec.
0°—5°	−0·051	−0·032	−0·061	−0·027	−0·003	−0·008	+0·043	+0·055	+0·071	+0·019	−0·003	−0·032
5—10	−0·035	−0·043	−0·024	−0·043	+0·008	+0·027	+0·035	0·000	+0·051	+0·043	+0·011	−0·019
10—15	−0·055	−0·043	−0·063	−0·039	+0·008	+0·019	+0·024	+0·032	+0·059	+0·047	+0·011	−0·024
15—20	−0·043	−0·047	−0·047	−0·032	+0·027	+0·027	+0·035	+0·066	+0·055	+0·043	−0·011	−0·008
20—25	−0·039	−0·056	−0·028	−0·056	+0·000	+0·038	+0·010	+0·053	+0·068	+0·081	−0·023	−0·025
25—30	−0·019	−0·043	−0·035	−0·068	+0·043	−0·002	+0·035	+0·082	+0·039	+0·066	−0·016	−0·011
30—36	−0·068	−0·049	0·000	−0·049	+0·030	+0·003	+0·055	−0·043	+0·019	+0·063	+0·020	−0·066
Mean	−0·044	−0·044	−0·035	−0·045	+0·016	+0·015	+0·034	+0·048	+0·061	+0·052	−0·002	−0·024

* In both of the Tables (7 & 8), which refer to the Southern Hemisphere, the seasons are taken in the same sense as in the Northern Hemisphere; viz., Winter—December to February; Summer—June to August.

(7) SOUTH ATLANTIC OCEAN (continued).

Latitude S	Winter	Spring	Summer	Autumn	Mean	Oscillation
0° – 5°	−0·036	−0·026	+0·031	+0·031	29·939	0·067
5 – 10	−0·030	−0·030	+0·023	+0·038	29·979	0·068
10 – 15	−0·038	−0·030	+0·027	+0 024	30·025	0·066
15 – 20	−0·038	−0·022	+0·038	+0·023	30·064	0·076
20 – 25	−0·042	−0·029	+0·031	+0·118	30·104	0·160
25 – 30	−0·033	−0·026	+0·031	+0·025	30·101	0·064
30 – 35	−0·062	−0·007	+0·034	+0·035	30·053	0·097

(8) INDIAN OCEAN.

Latitude S		Winter	Spring	Summer	Autumn	Mean	Oscillation
16° – 23°	19°·5	−0·019	−0·003	+0·021	+0·014	30·018	0·010
20 – 26	23	−0·067	−0·020	+0·071	+0·061	30·076	0·138
23 – 28	25·5	−0·102	−0·039	+0·114	+0·039	30·076	0·216
27 – 31	28	−0·106	−0·019	+0·098	+0·034	30·067	0·201
30 – 33	31·5	−0·131	−0·010	+0·151	+0·005	30·071	0·282
33 – 36	34	−0·072	+0·007	+0·055	+0·042	29·996	0·127

I append graphical representations of the rarefactions which are observed throughout the area where that phenomenon is most remarkable.

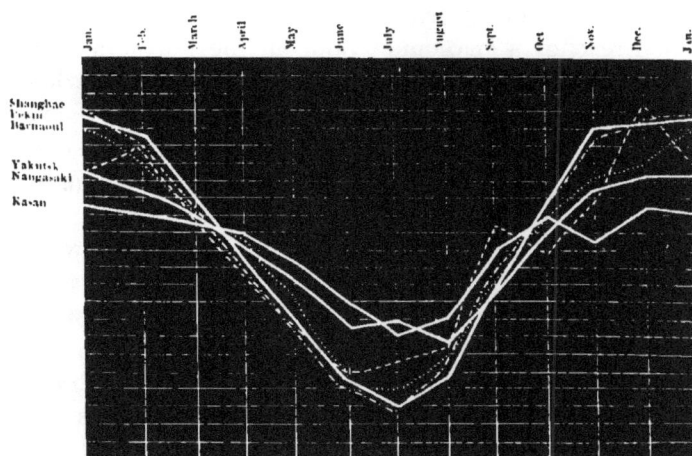

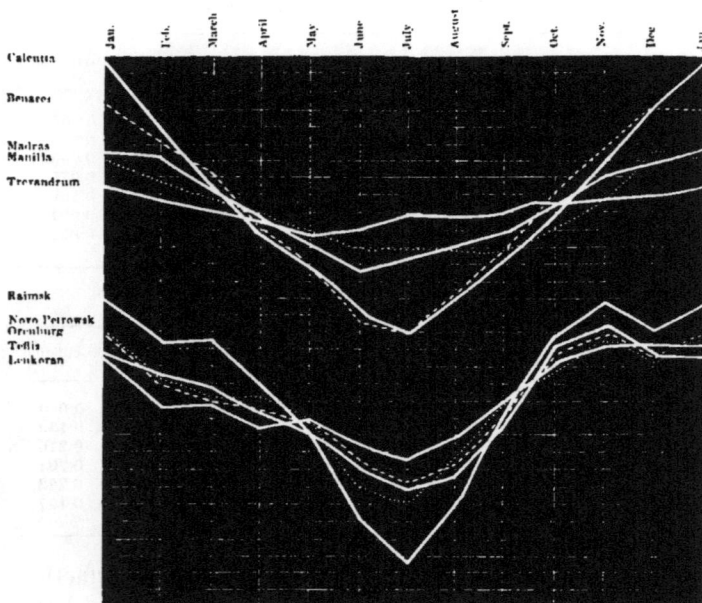

If we seek to determine the area in which the rarefaction reaches a maximum value in the northern hemisphere, we find it to exceed 0·35 in. over an area whose
southern limit passes from Barnaoul to the shores of the Sea
of Aral, then turns to the eastward so as to enclose the
whole plateau of the Gobi, and even to embrace Shanghae
and Pekin on the east coast of Asia, while Benares and
Calcutta are all but included within it. The area in which
the rarefaction is between 0·18 and 0·35 in. includes the
eastern side of the Ural range lying to the south of
Catharinenburg, the Kirghis Steppe, the Caspian Sea,
Persia, Afghanistan, Southern Arabia, the catchment
basin of the Indus, the valley of the Ganges, the
northern portion of Burmah, and the south of China. The

boundary turns northwards, between Canton and the Philippine Islands, and passes through Japan to the coast of Mantchouria.   The extreme limit of the area in the Pacific Ocean — *i.e.*, the point where the rarefaction ceases to be observable, is as yet undetermined.   The western limit passes from the neighbourhood of St. Petersburg to the south, so that the west coast of the Black Sea, Asia Minor, Syria, and Egypt, are included in it.   The point where this boundary meets the coast of Africa may be determined with considerable accuracy, inasmuch as Palermo and Tripoli lie outside of it, as may be seen from the following table : —

	TRIPOLI	PALERMO		
	Barom.	Barom.	Dry Air	Tension of Vapour
January	+ 0·122	+ 0·004	+ 0·158	0·314
February	− 0·021	+ 0·013	+ 0·161	0·320
March	− 0·086	− 0·036	+ 0·072	0·359
April	− 0·058	− 0·054	+ 0·024	0·390
May	− 0·068	− 0·014	+ 0·006	0·449
June	+ 0·029	+ 0·020	− 0·076	0·564
July	+ 0·004	+ 0·004	− 0·200	0·672
August	+ 0·020	+ 0·006	− 0·177	0·651
September	+ 0·019	+ 0·030	− 0·119	0·616
October	− 0·002	+ 0·018	− 0·054	0·541
November	− 0·014	+ 0·012	+ 0·070	0·410
December	+ 0·052	− 0·009	+ 0·126	0·333
Year	30·104	29·711	29·244	0·468

The stations in Algeria exhibit clearly the deflection of the barometrical curve, while those in the Azores and Canary Islands do not do so.   It is, therefore, probable that the desert of Sahara is included within the area of rarefaction, as Abyssinia has been proved to belong to it.

The distribution of atmospherical pressure does not agree directly with that of temperature.   In the month of

July the isothermal lines form a closed curve in Abyssinia
and Arabia, enclosing an area in which the temperature
reaches a maximum value. From this point the temperature decreases materially towards the district of maximum
rarefaction in northern Asia. As regards the western
boundary, the coincidence with the is-abnormals is much
closer, but even this does not hold good on the east coast
of Asia. If we remember that all the great deserts are
included within the area, we are referred so directly to
the action of the aqueous vapour that we are justified in
expecting to find the grounds of this phenomenon therein.

In spring, according as the sun's meridian altitude
increases, the temperature in Siberia rises very rapidly,
so that the difference of temperature between northern
Asia and Hindostan is materially decreased. It is still,
however, sufficient to preserve the current of air as NE.
Monsoon from higher to lower latitudes. After this period
the barometrical difference is considerably diminished.

In Barnaoul, the tension of aqueous vapour does not
exceed 0·18 in. from January to May, while in Calcutta it
rises 0·35 in. On the other hand the barometer falls 0·35 in.
in Calcutta during the same period, and in Barnaoul
nearly 0·44 in. By the month of June the total fall in
Barnaoul reaches 0·66 in., and in Calcutta only 0·49 in.
We see from this that the point of minimum atmospherical
pressure lies to the northward of that of maximum temperature, so that the current of air from the southward is
attracted not only to this latter point, but beyond it.
Another consequence of the rapid change of position of
the point of minimum pressure is that the SW. Monsoon
frequently sets in suddenly. Under other circumstances
we should only expect the NE. Monsoon to set in suddenly,
as it is a colder current than the other.

The long continuance of the SW. Monsoon is explained

by the fact, that this current, in its passage over the southern slopes of the table-land of Central Asia, parts with a great portion of the moisture which it contains in the copious precipitation of rain which there takes place, and consequently is insufficient to fill the vacuum which exists over that plateau. The monsoon is in nowise a phenomenon produced by the attraction of air from a colder towards a warmer point, analogous to the daily land and sea-breezes before described, as is still maintained by modern writers who follow in the steps of the old text-books. Its cause is to be sought for in the temperate rather than in the torrid zone.

The direction of the monsoons is thus accounted for. The SE. Trade-wind owes its motion from the east, to the fact of its passage from points which possess a slow to those which possess a more rapid velocity of rotation. As soon as this current crosses the equator, conditions, which are exactly opposite to the above, occur, and the wind becomes SW. In precisely the same way the NE. Trade-wind, on crossing the equator, becomes the NW. Monsoon of the southern hemisphere.

2. The West Monsoons of the Line.

This is a peculiar modification of the Trade-winds on the Atlantic Ocean, which is known on the coast of Guinea. Owing to the position of Upper Guinea, the SE. Trade-wind is attracted so far to the northward that rain-bringing winds from the SW. and WSW. are felt over the whole district up to the Cape Verde Islands. Between this deflected trade-wind and the ordinary trade, there is interposed a region of calms, which is marked on the maps of Halley, Dampier, and Mushenbroek, as the ' Rain ' or 'Thunder-sea,' in consequence of the frequency of tornadoes.

The observations which were carried on in the Danish Fort Christiansborg, on the Coast of Guinea, between the years 1829 and 1834, give a mean SW. direction of the wind, which continues unchanged throughout the year from 10 o'clock A.M. to 10 P.M. In the morning hours the direction is slightly north-westerly.

Similar conditions seem to exist in the Pacific Ocean between the Galapagos Islands and the coast of Mexico. Their explanation is the same as the foregoing, and is to be sought in the sudden change of direction of the coast of North America, which is at first northerly, and afterwards suddenly trends away to the north-west.

This is indicated on Dampier's chart. In the Gulf of Panama he gives easterly winds from September to March, and SSW. from March to September. In Captain Wilkes' 'Map of the World, showing the Extent and Direction of the Wind,' they are represented as spurs of the coast wind, called the 'Peruvian Monsoon;' and the Galapagos are placed on the boundary between unchanged Trade-wind and this altered current. Maury's Wind-chart (Plate xv. Part I. 8th edition) gives these winds as SE. and SW. Monsoon at the southern edge of the region of calms, which here, as in the Atlantic Ocean, extends far to the northward, and which English sailors call the 'Equatorial Doldrums.'

The coast of Venezuela must exert an influence the exact converse of this. In fact, the observations made at Cayenne from 1846 to 1852, show the mean direction of · the wind from December to April decidedly NE., and from April to November E. During the first-named five months, the proportion which the NE. wind bears to the E. is 669 : 78. In the remaining months it changes

regularly as follows : — May 59 : 68, June 44 : 105, July
18 : 140, August 8 : 163, September 16 : 164, October
33 : 131, November 56 : 101.   Simultaneously with this
the trade-wind near the coast becomes fresher, as it is
flowing towards the heated area before it.   The Spanish
sailors call these very strong Trade-winds at Carthagena
los brisotes de la Santa Martha, and *las brizas pardas*
in the Gulf of Mexico.   (Humboldt's *Reise* ' Travels,'
vol. i. p. 171).

3. THE LATERAL DEFLECTIONS OF THE TRADE-WIND.

The only modifications of the Trade-wind that we have
hitherto considered, are those produced by the relative
positions of land and water, considered with reference to
latitude.   If the line of separation between land and
water be a north and south, instead of an east and west
line, the result is the common phenomenon of lateral
deflections of the Trade-wind, the so-called ' Coast-winds.'
These winds are frequently intermitting, within the daily
period.   When a coast line is a north and south one, the
Trade-wind, in its passage along it, is attracted during the
daytime towards the land, which is at an elevated temperature, while at night it does not suffer this deflection,
whereas the so-called Land and Sea-winds alternate within
the daily period.   Similarly it may happen, that a lateral
deflection of the Trade-wind, of the above character, may
exist for a lengthened period, and so give rise to a sort of
monsoon.   The difference between such winds and the
true monsoons is, that in the latter case the winds have
directions opposite to each other, while in the former the
directions are inclined to each other at an angle which
is more or less obtuse.

According to Glass (History of the Canary Islands),
Africa exerts a very important influence on the direction

of the NE. Trade in the Atlantic Ocean.   The NE. wind
suffers a deflection towards the coast, which increases in
magnitude the nearer the islands, on which the observations are taken, lie to the mainland.   In sight of land the
wind is nearly N. (viz. N. by E.).   At Lanzerote and
Fuertaventura NNE., at Grand Canary NE., at Teneriffe
NE. by E., and lastly, at Palma slightly more easterly,
which direction it preserves over the whole Atlantic
Ocean.

These winds are stopped by the high islands of Canary,
Teneriffe, and Palma, so completely that, while they are
blowing strongly on the NE. side of the islands, the opposite side is in complete calm.   Glass gives the extent of
this shelter to seaward, as 20 to 25 nautical miles for
Grand Canary, 15 for Teneriffe, 10 for Gomera, and 30 for
Palma.   These calm island shadows are very dangerous
to navigation, as the high waves break on the edge of the
calm water as if on a shore, and produce a dangerous surf.

Dampier * gives as limits of this in-shore north-wind,
Cape Verde in lat. 14° and Cape Bojador in lat. 27°.   He
states that ' the dust-falls characterise the middle of the
district.'   'The ships being to the southward of Cape
Blanco, which lies in lat. about 21, they are sometimes so
troubled with the sand which the wind brings off shore
that they are scarce able to see one another.   Their decks
are all strewed with it, and their sails all red, as if they
were tanned with the sand that sticks to them, it being of
a reddish colour.'

Horsburgh (India Directory, vol. i. p. 11) states that
the dust atmosphere between the Cape Verde Islands and
the continent is a phenomenon which always exists, and
continues with NE. winds, and attributes it to the drift in

---

* Dampier, Voyages, vol. ii. part 3 (The General Trade-winds), p. 15.

of sand from the hot sandy desert which lies to the north-east. In fact, the phenomenon agrees perfectly with that observed on the Guinea Coast, with the NE. land-wind which sets in, especially in December and January, and is called there 'Harmattan.' 'A wind which produces unusual dryness, and fills the air with reddish dust so as not unfrequently to obscure the sun to such a degree that it may be looked at with the naked eye.' During the prevalence of this wind the tension of aqueous vapour at Christiansborg at 7 A.M. is 0·622 in., while with ordinary winds it is 0·846 in.; at noon 0·651 instead of 0·940. In the latter instance the ratio of the fractions of saturation is 0·734 to 0·915 ; in the former 0·473 and 0·739. (*Observationes Meteorologicæ in Guinea Factæ*, 1845, p. 50.) (Meteorological Observations made in Guinea.)

From reasons similar to the above, the SE. Trade in the South Atlantic Ocean is more southerly on the African coast, and more easterly on that of Brazil, than in the middle of the ocean.

The chain of the Andes endeavours to exert a similar influence, on a much larger scale, on the SE. Trade in its passage over the continent of South America, as the Canary Islands do on the NE. Trade in the Atlantic Ocean. We should expect to find a sheltered area on the lee-side of these high mountains along the coast of the Pacific Ocean. As in the case of the Canary Islands, where the waves break on the edge of the calm water ; so, in the district we are now describing, the Trade-wind, which is in violent motion, forces its way along the coast, and produces instead of a calm a SSW. wind, when the coast has a N. and S. direction ; SW. when its direction is from SW. to NE., and finally S. when the direction of the coast is from SSE. to NNW. On the coast of Peru these winds extend without any variation to a distance of from 140 to 150 nautical miles to seaward. They then pass gradually into the ordinary Trade-winds, and at a distance of 200 miles the true ESE. Trade is found.

We may naturally expect that the magnitude of the deflecting influence, which is exerted by the land on the wind which is prevalent at sea, will be dependent on the relation of the temperature of the water to that of the land; a relation which is not constant throughout the year. We may see from the form of the monthly isothermal lines, that the variation in this relation will not be very great, inasmuch as the cold water current, which flows along the coast of Peru, changes its temperature very slightly within the year. In addition to this, the temperature of the land varies between very narrow limits. On Wilkes' map these winds are called 'The Peruvian Monsoon,' in the same way as he calls the deflection of the SE. Trade-wind on the south coast of Africa 'the African Monsoon.'

Dampier, with much greater correctness, classes them with the Coast-winds, inserting them in an intermediate position between the permanent winds and the 'winds that shift,' which deserve the name of monsoons much better, and on the coast of Brazil frequently receive that name.

Wilkes says of them, 'Properly speaking, they are not monsoons, though they have been classed and considered as such. They do not interchange their direction periodically, like the others, but only veer through several points of the compass.' The district where the phenomenon is observed, extends from Cape St. Roque to the Island of St. Catherine. In this district NNE. winds prevail from September to February, and even, according to Dampier, up to March, during the period that the sun is at the south of the equator. During the same period, all portions of

the continents of the southern hemisphere, which extend
into the torrid zone, experience an atmospherical rarefaction, as is shown by the following table of barometrical
observations : —

	Port Jackson	Cape of Good Hope	Rio Janeiro	Montevideo	St. Jago
January	−0·145	−0·105	−0·098	−0·096	−0·052
February. . . .	−0·066	−0·105	−0·079	−0·097	−0·053
March	+0·036	−0·067	−0·047	−0·041	−0·035
April	−0·025	−0·033	−0·021	+0·039	−0·001
May	+0·035	+0·033	+0·068	+0·052	+0·006
June	+0·079	+0·093	+0·127	+0·051	+0·036
July	+0·170	+0·124	+0·135	+0·118	+0·018
August	+0·074	+0·111	+0·074	+0·079	+0·074
September . . .	−0·012	+0·062	+0·060	+0·009	+0·040
October	−0·002	+0·015	−0·028	−0·024	+0·032
November . . .	−0·058	−0·051	−0·090	−0·007	−0·031
December . . .	−0·091	−0·080	−0·099	−0·113	−0·033
Mean	29·514	30·036	29·843	29·955	28·075

If Brazil were to change places with the South Atlantic
Ocean, the result would be that the NE. Trade-wind of
the Atlantic Ocean would cross the line into the southern
hemisphere as a NW. Monsoon. In this case Brazil would
play the same part, with regard to this wind, as Australia
does with regard to the NE. Trade of the Indian Ocean.
Its more westerly position causes the NE. Trade to preserve
the same direction in the southern hemisphere, as it had
in the northern, instead of assuming one which would be
north-westerly, owing to the increasing difference in the
velocity of rotation. The effect of this difference is, however,
observable, for the wind, which is almost easterly in the
Gulf of Mexico, increases in intensity as '*los brisotes de
Santa Martha*' at Carthagena becomes NNE. at Cayenne,
and retains this direction along the coast of Brazil.
   The coast-winds of the east coast of Africa resemble

those which we have just been considering in so far as that,
when the sun is to the south of the equator, and the barometrical observations, at the Cape Mauritius and Bourbon,
exhibit a rarefaction, there is a prevalence of north-east
winds in the Mozambique channel and along the whole
coast of the continent, but without the permanence of the
true monsoon.   When the sun is to the north of the
equator the SE. Trade changes, in the neighbourhood of
the coast, into SSW. or south.   According to Capper
(p. 69), this NE. Monsoon commences at the Comoro
Islands in lat 10° S. in November, somewhat later than on
the Malabar coast, and extends to Cape Corrientes, *i.e.*
as far as the tropic of Capricorn.   The SW. Monsoon
begins in April, and lasts till November.   It is accompanied
in the Mozambique channel by clear weather, while the
weather is rainy during the NE. Monsoon.   At the change
of the one into the other in November, there are daily land
and sea-breezes.   The relative position and character of
the land appears to modify these winds variously.   As
regards direction they agree, on the whole, with the monsoons of the North Indian Ocean, but differ from them in
their contrast to each other with reference to the aqueous
vapour which they contain.   On Maury's chart they are
given as SW. and SE. winds, in contradiction to all observations that I have seen.

During the period that the sun is to the south of the
equator, Australia appears to exert an attractive action
on all sides on the direction of the winds which prevail
on the seas contiguous to it, and in fact, as far as the
observations, which have as yet been made, go, they
indicate a rarefaction of the air in that portion of the continent which is in the torrid zone.   The influence of its
northern coast on the monsoon has been already explained.
A similar deflecting action is also observable on the west

coast, as the SE. Trade, on approaching the coast, passes gradually through S. towards W.   On the east coast, Wilkes' chart gives easterly winds everywhere, and even ENE. on the northern portion of the continent.   In the case of Australia, as well as in that of the other continents which extend into the southern hemisphere, there is another circumstance to be taken into consideration, viz. that the period of greatest altitude of the sun in the southern hemisphere is also that of its greatest proximity to the earth.   However, the effect of this contrast of the two hemispheres is removed by the fact that, when we consider the total effect of the sun's position in the year, we find that during the longer period of the summer of the northern hemisphere, when the sun is at a greater distance from the earth, the amount of heat received by that portion of the earth is the same as that imparted to the southern hemisphere by the sun, which is at a less distance from the earth throughout the shorter summer of that hemisphere.   A direct consequence of this is that the direct radiation at the period of maximum altitude of the sun in the southern hemisphere is greater than it is at the same latitude to the north of the equator, during the period of summer in that district.

If, in conclusion, we collect all the circumstances of the winds of the torrid zone into one general view, considering their variations in various parts of that area, we obtain the following results : —

Permanent winds, properly so called, exist only at the centre of the two Trade-wind zones, *i.e.* at those places which continue between the limits of those zones during their annual motion northwards and southwards.   The strongest contrast to this is found at those places which do not pass out of the belt of calms during this motion.   The breadth of the Trade-wind zone is much greater than that of the belt of Calms, so that the instance last mentioned is only partially realised. At places situated close to the equator, the constant Trade-wind becomes an *intermittent* wind. If the extent of the motion of the Trade-zone be considerable, the station will be included in both the Trade-wind zones, so that as it is exposed to both the Trade-winds, the permanent wind above mentioned will be converted into an *alternating* wind. This periodical wind receives its developement as a true *Monsoon*, when the alternating winds receive, from the influence of the rotation of the earth, a direction opposite to each other, and this wind appears in its most complete form when the duration of the two Trade-winds is of equal extent, and the year is divided consequently into equal portions. The condition of opposite directions will not be fulfilled if the duration of the two Trade-winds be not equal, inasmuch as the difference of the velocity of rotation is dependent on the distance of the point of observation from the point where the atmospherical current commences. It may further be modified by lateral deflections which have a greater effect on one current than on the other, since the relation between the temperatures of the land and the water is not the same in winter as in summer. However, the Monsoons and the coast-winds are not the only classes of periodical winds. These winds appear in sub-tropical regions as well, but in exactly opposite directions. When the sun has its greatest altitude they are included in the Trade-wind zone, but as soon as that altitude begins to decrease, pass out of the zone, and are immediately included in the descending return Trade, whose direction is directly opposed to that of the preceding wind. These winds may be termed *sub-tropical*. This region forms the passage to the district of the *changeable* winds, in which the stations are included in each current in turn; but there is no definite period for this alternation. The two currents flow alongside of each other, in channels which are constantly changing. The name of upper and lower Trade-wind could not with any correctness be applied to them; and I have therefore invariably termed them in all my publications since the year 1827 the *Polar* and *Equatorial* currents.

Before we proceed to treat of these changeable winds, we must not omit to consider one omission which every-one will have noticed in the above explanation. This is, that we have considered the constant and periodical winds in their course over the different oceans, and in the vicinity of the coast. It is, however, clear that the general relations which call them forth are felt in the same way on land, but that there are so many disturbing causes, such as the unevenness of surface and other variations in its condition, that we cannot expect to find here the same simplicity as appears at sea. In the interior of continents we have to look to the commencements of the wet and dry seasons, to find the time of entrance of the station into either Trade-wind zone, apart from the constant action which the prevalent winds in the deserts exert, by the accumulations of sand which they leave behind them. The passage of this direction of the wind in the interior of the continents, which is constant at certain periods of the year, into the modifications of the same wind which are due to the coast, is as yet completely unknown, and, as far as I know, has never yet been discussed. It does not seem superfluous to indicate this as a very important subject of enquiry, as the fact that the Trade-wind charts are only published for the oceans has produced in the minds of many persons the strange idea that trade-winds are only felt at sea. This conception has been at last fully given up in the case of the Monsoons, which were long considered to be a kind of land and sea-winds ; although these latter disappear entirely at a distance from the coast, while the Monsoon does not part with the greater portion of the aqueous vapour which it contains until it comes in contact with the southern slopes of the table-land of Asia.   It appears suitable to introduce this remark here, as we shall establish the laws of the changeable winds on observations obtained at land stations, and must claim the right of extension of these laws to the ocean.

III.—The Changeable Winds.

ARISTOTLE, in the Politics, says that there are, properly speaking, only two forms of Constitutions, viz. those which are free, and those which are not free; in the same way as it is said of the winds, that there are really only two winds, viz. northerly and southerly, inasmuch as all other winds are merely occasional deviations from these two directions. Such a statement as the above seems less paradoxical on the coasts of the Mediterranean than it would in our latitudes, for the contrast of the 'Tramontane' and the 'Sirocco' is so clearly marked, at the exterior boundary of the Trade-wind area, that the former wind, which is the prolongation of the Trade-wind towards the north, was as well known to the Greeks, under the name of the 'Etesian' wind in summer, as the rain-winds from the opposite point in winter. We shall, however, proceed to prove that the above statement holds good for districts in which a dry season is not known, and this must be done by means of investigations of a special character.

Investigations of this character must be carried on in a twofold manner: we must prove, firstly, that the changes in the direction of the wind at any station may be explained by an alteration of two currents; and, secondly, that these currents, flowing simultaneously in parallel channels, are able to produce the contrasts of weather which are observed in succession at each station when the change of wind takes place.

As regards the relative duration of the two currents, it is evident that the equatorial must predominate over the polar, and consequently that the prevalent direction of the wind, in the temperate zone of the northern hemisphere, must be SW., of the southern hemisphere NW.   The grounds of this predominance are, firstly, that the equatorial current moves in a channel which is gradually contracting, the polar in one which is gradually widening ; secondly, that the air, when it returns from the poles to the equator, occupies less space than it did previously, owing to its reduction of temperature ; and thirdly, that it has lost by condensation a quantity of moisture, which was present in the form of vapour in the equatorial current, but flows back to the equator as water, having been precipitated in the form of rain.

On the whole, accordingly, we see that the currents flow in the temperate zone in parallel channels, while one flows above the other in the torrid zone.   We have described above (p. 39) the gradual descent of the upper current at the outer edge of the Trade-wind zone.   To this motion in parallel channels we owe the fact that whenever a relatively severe winter is observed, there exists, at the same time, a relatively mild one in an adjacent district.   This simultaneous compensation of a local excess and defect may be shown to lead at all seasons to a constant mean temperature, as I have clearly proved in the Investigations on the Non-periodical Variations in the Distribution of Heat on the Surface of the Globe (6 vols.).*

----

* *Untersuchungen über die nicht periodischen Aenderungen der Temperaturvertheilung auf der Oberfläche der Erde.*   Reimer, Berlin.

1. The Currents which alternately Displace each other at any Station.

The currents themselves may be distinguished from each other at any station by the properties explained below.   The equatorial current flows from a warm to a cold, the polar from a cold to a warm region.

The characteristic differences of these currents may always be traced to their *differences of temperature*, and to the *differences in the action exerted by the earth on them in their course.*

As regards *Pressure*, it is evident that, owing to the greater density of the cold air of the northerly current, the barometer will be high during its prevalence; and that, on the contrary, its level will be lower the greater the intensity of the southerly current, or, in other words, the greater the difference in latitude between the points at which the current commences and at which it ends.

As regards *Direction*, we have shown above, that the northerly current must, when it is well established, take a NE. and at last almost an easterly direction; the southerly current a SW., and at last almost a westerly direction. Most winds are liars, as they do not come from the regions from which they appear to flow.   The ENE., the NE., and the NNE. are much more truly northerly winds than the N. itself, and in the same way the WSW., the SW., and the SSW. much more truly southerly than the S. itself.

As regards *Intensity* it is easy to see that that of the northerly current must gradually diminish on its progress; that of the southerly gradually increase.   If the earth were a right cylinder revolving about its axis, the velocity of rotation would be constant at all parallels of latitude, and the space between two meridians would be the same all over its surface.   Under such circumstances, the direction and intensity of the northerly and southerly currents would remain constant throughout their course. Owing to the spheroidical form of the earth, the meridians diverge from each other between the pole and the equator, and consequently the channel of the northerly current is constantly dilating the farther it gets to the southward, while that of the southerly current is closing in a corresponding manner the farther it gets to the northward. The intensity of the northerly current, accordingly, diminishes as it becomes easterly, while that of the southerly current increases as it becomes westerly.

As regards *Moisture*, it is evident that the northerly current must be the drier, the southerly the damper, as well relatively as absolutely. Inasmuch as they start from points whose temperatures are very different; under circumstances, in other respects identical, the absolute amount of aqueous vapour contained in the northerly current must be less than that in the southerly. The warm equatorial air flowing rapidly towards the pole enters higher and higher latitudes; accordingly, its capacity for retaining moisture is gradually diminished, owing to the cooling action of the surface of the earth upon it, and it must therefore gradually approach its point of saturation during its progress. On the other hand, the colder air of the slowly moving northerly current reaches lower and lower latitudes, its capacity for moisture is constantly rising, and it therefore absorbs moisture, instead of losing it in the form of rain. We see accordingly that while the southerly current loses the aqueous vapour which it brought with it in constantly recurring discharges of rain, there is always a clear sky during the prevalence of the northerly current, as there is during the north-east Trade-wind.

As regards the *Change of Direction* of the wind and the

ratio of the number of the individual winds, these points depend on very various circumstances. If, during the whole period of its prevalence, which is often for many weeks at a time, the origin of the northerly current were always equidistant from the point of observation, and further, were its intensity always the same, the direction of the vane would remain unchanged, *i. e.* as soon as the current had assumed the deviation which is due to the difference of velocities of rotation of the two points, and the other elements which exert an influence on it. If, however, the current, which is almost easterly, meets with an obstacle at a point situated to the south of the point of observation, and is prevented from converting itself, in its further progress, into a true east wind, the air composing that current will enter into the condition of relative rest to the earth in its rotation, and accordingly become a north wind at that point. This will cause the wind at the point of observation to become more northerly, so that the vane will oscillate slowly between NNE. and ENE. At different times the distance of the point of origin of this current from the point of observation will vary; and, further, the friction which is exerted by the surface of the earth on the lower strata of a current, will be greater when it is flowing slowly than when it is in more rapid motion; so that the easterly deviation will be more considerable. We may therefore expect to find that there is not much difference, as regards number, between NE. and E.; and the maximum will fall accordingly, sometimes on the one, sometimes on the other point of the compass. From considerations similar to the above it is easy to see that the quick-flowing southerly current, which is SW. when it has attained permanency, will exhibit, in general, an oscillation between SW. and W.; that the number of times that the wind will blow from these points will be very great, and that, if we take all these fluctuations into account, we shall find a majority of direct shiftings, which cannot be very considerable, and must be proportionably less so the shorter the intervals are which intervene between the periods of observation. South and north winds, in the sense of the Law of Gyration, as they come immediately *before* a maximum, will be, in general, of rare occurrence, if local conditions, such as the direction of a line of coast, or that of a valley, do not confine the currents to these points. According to the above we should expect also the SE. and NW. winds to be rare winds. This is the case with the SE.,* but by no means with the NW., especially in the west of Europe.† Its

* A remarkable exception to this statement is found in the Steppe district of the south of Russia, in which tract of country, extending from the north coast of the Caspian to that of the Black Sea, and inland into the Steppe, the mean direction of the wind is south-easterly, according to the investigations of Wesselovski.

† From the observations of Mr. F. J. Foot, made at Ballyvaughan, on the west coast of Clare, in the year 1861, Prof. Haughton (*Journal of the Royal Dublin Society*, No. xxv.) has calculated the following table of the frequency of the wind (in percentages), referred to eight points of the compass, for that station, and compared it with the results obtained by Dr. Lloyd for the whole of Ireland in 1851 : —

	Co. Clare 1861	Ireland 1851
N.	13·9	10·8
NE.	4·9	7·3
E.	7·7	6·9
SE.	7·0	9·2
S.	19·2	13·6
SW.	12·3	20·2
W.	24·4	17·0
NW.	10·3	15·0
	100·0	100·0
Resultant	W. 16° 26′ S.	W. 22° 30′ S.

The extent to which the NW. wind prevails in Ireland is clearly shown by this table. (*Trans.*)

prevalence in summer is easily seen to be due to the position of the ocean with regard to the continent. The reason that the polar current, in winter, frequently commences as NW. is explained by the course of the isothermal lines, which, as I have shown at other times, must owe their position to other causes besides currents of wind. The low temperature of North America produces cold air-currents over the warmer area of the Atlantic Ocean, which meet the SW. wind of the southerly current, and, if they displace it, assume a NE. direction. In this case they probably leave the SW. on their eastern side. This latter wind flows on towards the pole, and may perhaps burst through the northerly current at a higher latitude, at which place the two currents will lie as extremes beside each other.

As regards the *mutual displacement* of the currents, I have ascertained from observations which have been carried on at Königsberg for a long series of years, that the southerly current displaces the northerly in the upper strata of the atmosphere, before that it does so in the lower strata ; while the displacement of the southerly by the northerly takes place at first in the lower and afterwards in the upper strata. A statement such as the above cannot be submitted to an accurate numerical examination, and must therefore be, in the main, founded on the assertions of different observers, so that the probability of its truth will depend on the mutual agreement of their testimony. When we come to direct observations of the appearance of the sky, there is no one on whom greater reliance may be placed than on Howard. In his *Climate of London* he says (p. 127) —

If after sultry damp weather, and a gradual accumulation of thunderclouds with electrical explosions, I observe the fall of small lumps of ice from the clouds; subsequently to this, violent hail, and at last rain; and if I find that a cold W. or N. wind ensues afterwards, I am justified in assuming that this latter wind has acted suddenly and in a decisive way in mass, as a cold body, on the warm air in which I was before the storm began. If, on the contrary, after a dry cold NE. wind the sky becomes overcast and the first drops of rain are warm to the feel, and if after a heavy shower the air below becomes warm and mild, I may conclude with as much certainty that the southerly wind has displaced the northerly, having commenced first in the upper atmosphere, and in displacing it has lost a portion of its moisture by condensation.

Apart from empirical proofs, like the above, it appears to me that this mode of displacement must necessarily follow from theoretical principles. Inasmuch as the northerly current flows between meridians which are constantly diverging from each other, the farther it has travelled, and the greater the easterly deflection which has been imparted to it by the motion of the earth on its axis, the more will the current dilate, and the total weight of the atmospherical column will be proportionably diminished. This is the reason of the fact that in winter — at which season all these conditions appear in their simplest forms — the pole of cold of the wind-rose lies much farther to the eastward than the maximum of the barometer. The temperature of the northern current is lower in proportion as its point of origin lies farther north. Under the same circumstances, the air composing that current is specifically heavier, and its barometrical pressure greater, if a cause of rarefaction does not at the same time arise from its having travelled to a long distance. Such a cause exists in the fact of its motion in a channel which is continually widening. Air, when it is expanding, falls in temperature, instead of rising; so that during the gradual lateral extension of the current the temperature will continue to decrease. The rarefaction of the air militates against the rising of the barometer, which ought to result from the change of temperature, and eventually causes the level of that instrument to fall. This shows us that the maximum of pressure will precede the minimum of temperature. In the case of the southerly current, which flows in a continually contracting channel, the lower air will have a constant tendency to rise, so that the diminution of density as well as the decrease of temperature in the southern current will be less considerable. If now the southerly current meets the northerly at a certain distance above the earth's surface, the elasticity of the air of the former will exceed that of the latter; and the southerly current will force its way into the northerly. It follows from the considerations just cited, that such a displacement can only take place when the northerly current has travelled for a long distance, and has been deflected to the eastward in a great degree, or in other words —

The southerly current can only displace the northerly current in the higher regions of the atmosphere, when the latter has become almost an east wind. This displacement must therefore appear as a shift in the direction E., SE., S., &c. Consequently, the wind hardly ever goes back in this quadrant of the compass.

If the southerly current has attained a considerable intensity, it may blow directly opposite to the northerly current. In this case the currents stop each other's way, a phenomenon which we shall examine more closely hereafter. We have now to explain how the shift takes place in the lower strata of the atmosphere.

The mixture of the warm moist air of the southern current with that of the northern, which is cold and dry,

gives rise to the conditions necessary for the precipitation
of aqueous vapour. The first form of cloud which appears is the Cirrus in long streaks. If fine spiculæ of
ice fall from this, they will reach the surface of the earth
with the velocity of the southern current, which, compared with that of the northerly current beneath, is very
considerable. The first drops may fall to the earth either
in the solid or the liquid form, or may be evaporated
before they reach the ground; but in either case they
will exert a certain action on the strata of the atmosphere
through which they pass. In consequence of this action
the wind will become first due east, then ESE., and thus
shift gradually until it is south, from whence it will soon
be brought to SW. by the rotation of the earth. In this
way we have deduced the phenomena of the east side of
the wind-rose, and accounted to ourselves for the facts that
precipitation seldom takes place with north-easterly or
easterly winds; that rain begins to fall when the wind is
to the south of east; that the NE. wind extends through
the whole of the atmosphere; and lastly, that when the
wind is E. or SE., its direction in the upper strata is more
southerly than in the lower. We now turn to the west-
side.

If the action of the northerly current on the southerly
never took place until the latter had reached the W.
point, the phenomena between S. and W. would exhibit
the characteristic results of the predominance of the
southerly current, while those between W. and N. would
show the effects produced by the displacement of this
latter by the northerly current. This, however, is not
always the case, as the northerly current often commences to blow when the wind is at SSW., SW., or
WSW., and therefore the phenomena between S. and W.
may exhibit a twofold character. It is self-evident that

the cold and dense northerly current will first appear in the lower strata, and that the motion of the particles of the northern current towards the south, and of the southern current towards the north-east, will compound an intermediate resultant motion which will become more and more northerly, through north-west from west, the more the northern current prevails. The difference in density of the air in the two currents is very considerable in the lower strata, and decreases as the distance from the earth increases, so that the change will take place very rapidly at the surface of the earth, and will be accelerated by differences of temperature between the two currents. The barometrical indications are affected chiefly by the lower strata, and hence we obtain the following rule :—

In oscillations of the barometer, the front of a wave is steeper than the back, or, more accurately, *the warm and light air is displaced by that which is cold and heavy on the west side, more rapidly than the cold and heavy air by that which is warm and light on the east side.*

I now proceed to give proofs of this mutual displacement of the currents, and in these proofs we may clearly trace the order of sequence which I have termed the LAW OF GYRATION.

2. THE REGULAR SHIFTING OF THE WIND IN THE NORTHERN HEMISPHERE.

$\rightarrow$ S., SW., W., NW., N., NE., E., SE., S. $\leftarrow$

We read in the Book of Ecclesiastes, chap. i. 5, 6. 'The sun also ariseth, and the sun goeth down, and hasteth to his place where he arose. The wind goeth toward the south, and turneth about unto the north; it whirleth about continually, and the wind returneth again according to his circuits.' I do not venture to express an opinion as to whether these words are intended to indicate the Law of Gyration, or only to denote the change as such.

1. ARISTOTLE: * αἱ δὲ περιστάσεις τῶν ἀνέμων καταπαυομένων εἰς τοὺς ἐχομένους γίνονται κατὰ τὴν τοῦ ἡλίου μετάστασιν. 'The changes of the winds, when they are dying away, towards those which succeed them, take place according to the motion of the sun.'

2. THEOPHRASTUS:† "Οταν δὲ μὴ ὑπ' ἀλλήλων διαλύωνται τὰ πνεύματα, ἀλλ' αὐτὰ καταμαρανθῶσι, μεταβάλλουσιν εἰς τοὺς ἐχομένους ἐπὶ δεξιὰ ὥσπερ ἡ τοῦ ἡλίου ἔχει φορά. 'When a wind is not dispersed by another, but dies away gradually, the shift to that which succeeds it takes place from left to right, in the same direction as the motion of the sun.'

3. PLINY:‡ Omnes venti vicibus suis spirant majore ex parte, aut ut contrarius desinenti incipiat. Quum proxime cadentibus surgunt, a lœvo latere in dextrum, ut Sol, ambiunt. 'All winds blow for the most part in their turns, or so that a wind should be succeeded by one from the opposite quarter. If they commence nearly from the same point as that which has just ceased, the shift takes place from left to right in the direction of the motion of the sun.'

4. LORD BACON, *England*, 1600.§ 'When the wind changes conformably to the motion of the sun — that is, from east to south, from south to west, from west to north, and from north to east — it seldom goes back; or if it does 'tis only for a short time: but if it moves in a contrary direction, viz. from east to north, from north to west, from west to south, and from south to east, it generally returns to the former point, at least before it has gone through the whole circle.

'If the south wind begins to blow for two or three days, the

---

* Meteorology, ii. 6, 18; Problema, 28, 31, p. 943.
† Περὶ σημείων ὑδάτων καὶ πνευμάτων (Concerning the signs of the rains and the winds), sect. 10, op. ed. Schneider, p. 293.
‡ Natural History, vol. ii. p. 48.
§ Bacon's Draught for the Particular History of the Wind, sect. xii. Philosophical Works by Shaw. London, 1733, 4to. vol. iii. p. 476.

north wind will sometimes blow suddenly after it: but if the north wind blows for the same number of days, the south wind will not rise until after the east has blown awhile.'

5. MARIOTTE, *France*, about 1700.[*] 'When the north and north-east winds cease, the east generally succeeds, and then follow the south and south-west winds. The south and south-west winds generally succeed the east in the temperate zones, and especially in France. In France the winds usually shift from east through south to south-west, then to the west, to the north and north-east, and seldom make an entire circuit in a contrary direction.'

6. STURM, *Germany*, 1722.[†] 'However, the irregularity of the changes of the wind is not altogether without rule. From observations extending over a period of many years, and from those which have been made now at the time that we are writing, we find that there is a certain periodical revolution of the wind,—viz. that most frequently as a general rule, after the west wind, the north wind blows; that this is followed by the east wind, after which the south wind makes its appearance, and this in its turn gradually passes over into the west wind. The intermediate points of the compass are also observed and the rotation is very rarely reversed, for the wind (if it by chance has returned to the south from the west) hardly ever passes in that direction beyond the east. Thus much is wanting to complete a full revolution in that direction, whereas full revolutions in the opposite direction are completed frequently, even several times in the course of one month. On this principle we have found a method, without further assistance, to perceive coming changes of weather at least for the next ensuing days, and even to foretell them with considerable accuracy. The whole of this we have found to be confirmed by repeated experiments.'

7. TOALDO, *Italy*, 1774.[‡] 'In fact, if there be no obstacle, the winds go the round of the horizon with the sun.'

---

[*] *De la Nature de l'Air* (Concerning the Nature of the Air), p. 160.

[†] *Physica Electiva sive Hypothetica*, tom. ii. p. 1206.

[‡] *La Meteorologia applicata all' Agricultura* (Meteorology applied to Agriculture), p. 62.

8. Poitevin, *Southern France.** 'When the winds have blown from the S. and SE. with violence and brought rain with them, they run through the SW. and W. points, and terminate with NW., which last brings back fine weather.

'The N. and NE. winds often pass over the E. and are succeeded by sea-winds (S. and SE.). It is very seldom that the N. winds veer directly to NW.; however, this sometimes takes place : in general they traverse the horizon passing by the E.'

9. Kant.† 'In our northern hemisphere the winds usually, when they change from N. towards NE., traverse the whole circle from left to right in the following order: *i. e.* first to E., then to S., and then to W. Those winds which change in the opposite direction, as from N. to W. and so forth, seldom complete a whole revolution.'

10. Romme, *Northern Temperate Zone of the Atlantic Ocean.*‡ 'According to an English captain in the East India Company's service, the dominant winds from the parallel of 30° N. to the frigid zone are in this sea from W. or WSW. He remarks further, that a high N. or NW. wind, which ends with a calm, is followed by a S. wind, which brings rain, and this, if it rise high, shifts to W., NW., and N. If these latter winds become violent, they sometimes shift to NE. and blow for several days together, or else end in a calm which is followed by a S. wind. If this latter become rather westerly it is accompanied by rainy weather and squalls, and it often comes back to the S. during the rain.'

11. Lampadius, *Freiberg in Saxony,* 1806.§ 'How exceedingly changeable are the winds in Germany! Yet I have remarked a kind of periodical movement in them, which is as follows. I suppose the wind to blow from the south, with a clear sky. The barometer falls, the weather becomes thick, rain follows. In the meantime the wind becomes westerly. The rain still continues, and the barometer rises. The wind changes to the NW.

---

* *Climat de Montpellier*, p. 65.   † Physical Geography, 1802, p. 282.

‡ *Tableaux des Vents, des Marées et des Courants* (Description of the Winds, Tides, and Currents), vol. i. p. 56.

§ *Systematischer Grundriss der Atmosphärologie* (Systematical Sketch of Atmospherology), p. 189.

Partial rain ensues. It grows colder. The barometer still rises, and the wind becomes N. and NE. The barometer is now at the highest point, the sky is serene, and the severest cold possible at the season prevails. The wind veers to the E. and the barometer falls a little, but the weather as yet remains clear. The wind veers to the SE., the barometer still falls. The temperature increases again. The wind then changes to the S., and the temperature reaches the highest degree agreeable to the season: the barometer falls, and we come back again to the first point. There are annually several such periods in every season. Sometimes it will take several weeks to complete this rotation of the wind through the whole compass; sometimes but a few days. The wind will very rarely change in a direction contrary to that above mentioned. In general, all changes from the left to the right of the horizon are more common with us, and a S. wind is the least frequent. There certainly exists a primary cause of this, which is, however, concealed by many casualties.'

Lampadius has, however, gone beyond this excellent description of the phenomenon. As Sturm had done before him, he has founded rules for prophesying the weather upon the supposed correctness of this law, and in his 'Contributions to Atmospherology' has examined to what degree they may be depended upon.

12. Dove, *Königsberg*. 'In the course of my observations at Königsberg, which commenced in September, 1826, I compared the direction of the wind with the indications given by the barometer, and I at once observed a remarkable phenomenon, an example of which is given in the succeeding table (p. 92), the heights being reduced to their values at 54°·5 F.

From this we see that while the barometer described an undulation, the wind veered round the whole compass. I have since observed rotations in the same direction, viz., S., W., N., E., S., at all seasons, but they are most striking during the winter. If a SW. wind blows with increasing violence, and finally prevails, it raises the temperature above the freezing point, and therefore snow cannot fall, but it rains, and the barometer arrives at its lowest state. The wind now shifts towards the W. and the thick

KÖNIGSBERG, Barometer and Vane.

Day	8 A.M.	Noon	10 P.M.	Wind	Appearance of the Sky
	Inches	Inches	Inches		
September 25 .	29·829	29·827	30·069	W.	Cumuli
26 .	30·247	30·311	30·392	NW.	Overcast
27 .	30·441	30·444	30·386	NE.	Clear
28 .	30·371	30·344	30·307	E.	—
29 .	30·264	30·229	30·258	E.	—
30 .	30·293	30·256	30·246	E.	—
October 1 .	30·229	30·223	30·217	ESE.	—
2 .	30·244	30·223	30·199	SE.	—
3 .	30·156	30·090	29·972	S.	Cirri
4 .	29·886	29·902	29·816	S.	Slightly overcast
5 .	29·798	29·786	29·781	S.	Rain
6 .	29·885	29·971	30·131	W.	Overcast

flakes of snow prove the approach of a colder wind,— a fact which is also indicated by the quickly rising barometer, the vane, and the thermometer.   If the wind is N. the sky grows serene, if NE. the maximum of cold and of the barometer is observed.   The latter however gradually begins to fall, and fine cirri show by the direction of the streaks at their commencement, the S. wind which had set in and which the barometer already indicates, although the vane feels nothing of it, and still points steadily to the E.   The southerly wind, however, descending, with gradually increasing force, displaces the E. wind : simultaneously with a decided fall of the barometer the vane indicates SE., the sky becomes gradually more and more overcast, and as the temperature rises and the wind shifts through SE. and S. to SW., the snow turns to rain.   There is now a recommencement, and the fall of rain on the east side is separated in a highly characteristic manner from that on the west side by a brief period of clear weather.   Once acquainted with the phenomenon, when it appeared in its most evident form, it was easy for me again to recognise it in its more irregular changes : nay, to class these latter as frequent resilitions, more usual on the west side.   Hence, therefore, I ascertained that, in this country at least, all winds are whirlwinds on a great scale (I have observed rotations of from 1 to 22 days), and that the rotation of these whirlwinds is, on an average, always in the same direction.'

13. HILDRETH, *North America, Marietta, Ohio.*[*] 'The winds perform a regular rotation in the direction SW., W., NW., N., NE., E., SE., S., never in the opposite direction; this fact is especially observable in spring and autumn.'

14. DUDEN, *North America.*[†] ' In the State of Missouri, the wind, in constant repetition, traverses within from ten to twenty days every quarter of the horizon, and always in the following order, going from E. through S. to W. and through N. towards E.' Duden adds that he had never observed a complete revolution in the opposite direction.

15. SCHÜBLER, *Germany,* says: 'The rotation of the winds in Germany takes place more frequently in the order of S. through SW., W., NW., N., NE., E., and SE., than in the opposite order of S., through SE., E., NE.,' &c. &c.

16. *Siberia.* According to a verbal communication from Baron von Humboldt, the regular rotation of the wind with the sun is known to the natives of that country.

17. VON WRANGEL, *Sitka.*[‡] ' In New Archangel the prevalent winds are SE. and SW.   If the wind shift from S. towards SW. and W., it is accompanied by violent squalls, and thunder-storms are frequently observed, especially towards the end of autumn (November) and in winter, never, however, in summer.   If the wind veer round from W. to NW., the weather clears, and a long continuance of fine weather in Sitka is always accompanied by NW. wind.   If it veers from NW. through N. to NE. we have again violent squalls, which sometimes last for some time. If it go further towards E. and on to SE., rain invariably follows, with a continuance of damp and cloudy weather.   This condition of the weather is especially prevalent when the S. wind veers backwards towards SE.   The barometer falls with SE. and NE. winds, it rises with SW. and NW. winds.'

Herr von Baer adds: ' It seems to me very interesting that Herr von Wrangel noticed at Sitka the normal rotation of the winds from N. through E. to S., and termed the opposite

---

[*] Silliman, American Journal, 20, p. 127.
[†] Journey to the Western States of America, p. 200.
[‡] *Bulletin de l'Acad. de St. Pétersbourg,* 1838, Dec., p. 17.

rotation retrograde, at the same time as Professor Dove in
Europe was propounding the Law of Gyration of the winds.'

18. DUPUIS DE MACONET, *Lyons.** 'Whenever the N. wind
gives way to the S. wind, the change always takes place through
E.; when it returns from S. to N., the change almost always
takes place through W.'

19. DRAKE, *Cincinnati,* 1850.† 'Our winds, as is generally
the case with winds in the northern hemisphere, veer from one
point of the horizon to another in a regular order. Every observer in the centre or southern portion of the valley is convinced of the correctness of this statement. The usual direction
of the change is from left to right when the observer turns his
face to the S., and the same when he turns it to the N. Hence
a SE. wind becomes successively S., SW., W., NW., N., NE.,
without a calm having intervened. A change in the opposite
direction never takes place without a calm following the N.
wind; and if in this case a S. wind ensues, this is in reality a
commencement of the rotation afresh. An easterly wind seldom
takes a true E. course, but is converted into SE., and the NW.
wind seldom veers round as far as E. Although the wind frequently blows from the three points S., W., and NW. in succession, it seldom reaches E. It usually begins as SE., and before
it reaches this point from N. it has generally totally died away.'

20. KANE, *Smith Sound, Renselaer Harbour.*‡ 'Dove's Law of
Gyration does not well admit of confirmation, owing to the
great prevalence of calms, and also to the rare occurrence and
slight intensity of the winds between S. and NW. Nevertheless
we may observe a tendency in the case of ENE., NNE., and
NNW. winds to shift in accordance with the law, in the proportion of three instances confirmatory to one contrary, *i.e.* with a
relative probability of ¾ in favour of the law.'

We now pass on to the direct enumeration of the
changes of the vane.

---

* *Ann. des Sciences Phys. et Nat. de la Société d'Agriculture de Lyon,* vol. i.
p. 493.

† Principal Diseases of the Interior Valley of N. America, p. 572.

‡ Kane's Meteorological Observations in the Arctic Seas, p. 81.

When I published my first investigations relative to the Law of Gyration in the year 1827, I stated my reasons for preferring the indirect evidence, which I shall indicate below, given by the variations of the barometer, thermometer, and hygrometer, to the direct evidence given by a simple enumeration of the number of times in which the change of the wind took place in accordance with the Law of Gyration or against it. Buys Ballot agrees with me in this view (Pogg. Ann. lxviii. p. 417), and he considers these proofs as more conclusive in proportion as they appear to be concealed by those changes.

If we trust to a simple enumeration of the number of changes of wind it is unavoidably the case that all changes of wind through an arc greater than 180° are considered as contrary to the law. This shows us accordingly that the hours of observation may be so arranged that every such calculation must lead to a false interpretation of the result. If the interval between the hours of observation exceed the mean duration of a half revolution, an actual confirmation of the law will be considered as a retrograde rotation, *i.e.* as a contradiction to the law. Before, therefore, we undertake such a calculation we must determine the average length of time required for the accomplishment of a full revolution. According to my experience, this can only be determined by direct observations made with registering instruments, or by comparing the results of observations made at shorter intervals with those made at longer ones. In the case when the vane oscillates backwards and forwards, the answer to the question is not to be sought for in the absolute number of times that it moves in one direction or the other, but in the relative lengths of the arcs to which the sum of these motions in each direction extend.

We cannot therefore wonder that Schouw,* who, without taking notice of this important fact, investigated 1100 changes of the wind observed by Neuber at Apenrade, found 559 of them to be in the direction S.W.N.E., and 457 in the opposite direction. We should wonder that he obtained so decisive a majority in favour of the law, since all fluctuations were taken into consideration.

Karlsruhe. Eisenlohr,† by an examination of 46,665 observations, extending over a period of forty-three years, finds the following proportion of the number of rotations in the direction S.W.N.E., to the number in the opposite direction:

Rotations	Winter	Spring	Summer	Autumn	Year
Of 180°	1·57759	1·75439	1·41451	1·36965	1·51807
135	1·04196	1·05858	1·03462	1·12883	1·06211
90	1·05479	0·98524	1·13167	1·05851	1·05869
45	1·00224	0·97302	0·95801	0·99628	0·98030
Total . .	1·09877	1·10024	1·07189	1·09142	1·08881

From this we see that the greater the arc through which the rotation takes place, the more decided is the predominance of the direct rotations over those which are retrograde; whence we see clearly that the smaller fluctuations are merely oscillations of the wind. Apart from the magnitude of the arc of rotation, this predominance is observable not only in the mean for the whole year, but also in those for the different seasons.

In order to exhibit at one glance the result of this investiga-

* *Collectanea Meteorologica sub auspiciis Societatis Danicæ edita* (Meteorological Observations collected and published under the auspices of the Danish Society), i. 1829.

† *Untersuchungen über den Einfluss des Windes auf den Barometerstand, die Temperatur, die Bewölkung des Himmels und die verschiedenen Meteore zu Karlsruhe.* (Investigations relative to the influence of the Wind on the level of the Barometer, the Temperature, the Clouds, and the various Meteorological Phenomena at Carlsruhe). Heidelberg, 1837.

tion, without the least intention of wishing to give a determination of the rate of rotation, I have calculated, by means of Lambert's formula, the mean direction given by all the observations taken before each wind, and also that given by those after it, and subtracted the latter from the former. The negative sign indicates a rotation in a direction contrary to the law.

	Winter	Spring	Summer	Autumn	Year
N.	5° 6′	10° 0′	8° 40′	7° 20′	8° 5′
NE.	1 24	2 34	0 27	2 25	...
E.	1 15	4 52	7 1	2 39	8 3
SE.	22 53	13 5	16 6	21 21	12 18
S.	0 19	2 42	4 42	−1 4	1 10
SW.	0 51	2 4	0 46	1 3	1 8
W.	1 24	1 5	−0 13	1 26	0 41
NW.	11 23	11 12	7 40	10 45	9 50

The hours of observation being 7, 2, 9, the changes of direction are referred to periods of seven or ten hours.

Berlin. EMSMANN* has counted the rotations at Berlin in the years 1831—1835. He finds 347·2 direct, and 277·8 retrograde, so that the excess of the former over the latter is 69·4. He further finds that the yearly average of complete direct revolutions without any check whatever is 12, of retrograde 3.

Gnadenfeld near Kosel in Silesia. KÖLBING† finds in one year 10 complete direct revolutions; 1 of 3 days' duration, 4 of 4 days', 1 of 7, and 1 of 19 days'. In the following years, 1844 and 1845‡, when the observations were at times interrupted, he finds 21 complete direct rotations, of durations varying from 16 hours to 11 days, and not a single complete retrograde revolution.

Lenzburg in Switzerland. HOFMEISTER§ finds for 6 years the proportion of the retrograde rotations to those which are direct, for the various arcs, to be as follows:—

* *Untersuchungen über die Windverhältnisse in Berlin* (Investigations on the Circumstances of the Wind at Berlin). Frankfort-on-the-Oder, 1839.

† *Meteorologische Beobachtungen* (Meteorological Observations). Poggendorff's Annalen, lxii. p. 373.

‡ Pogg. Ann. lxxi. p. 309.

§ *Neue Denkschriften der allgemeinen schweizerischen Gesellschaft* (New Memoirs of the General Society of Switzerland), 10, p. 54.

Arc 22½° . . . 1 : 1　　　　Arc 112½° . . . 1 : 1·0769

45 . . . . 1 : 1·2754　　　　135 . . . . 1 : 1·1944

67½ . . . 1 : 1·4839　　　　157½ . . . 1 : 0·7879

90 . . . . 1 : 1·1379

Netherlands. BUYS BALLOT'S[*] calculations embrace 34 years in 4 groups, viz. 1730—1737; 1737—1743; 1749—1758; 1760—1769. The observations were instituted on 3 occasions by Muschenbroek. The calculations were made in the following manner :— If the wind were originally W. this was noted at the top of the column ; if at the next observation it was NW., a vertical stroke was made in the column between W. and NW. If NE. followed after NW. two such strokes were made, one between NW. and N., and one between N. and NE. If N. followed NE. a horizontal stroke was made between N. and NE. If SW. followed N. three such strokes were made; viz., between SW. and W., W. and NW., and NW. and N. In this way all direct rotations with their intermediate stages were marked by vertical, all retrograde by horizontal strokes. If the directions given by two successive observations were diametrically opposite, this was noted in a column by itself as ' leap,' since it could not be determined in which direction the rotation had taken place. The following table contains the results of this calculation ; (−) signs, as usual, indicate abnormal gyrations.

WINTER

	1730—1737	1737—1743	1749—1758	1760—1769	Sum
S.— SW.	2	27	33	17	79
SW.— W.	0	28	32	14	74
W.— NW.	0	29	27	13	69
NW.— N.	− 4	30	28	8	62
N.— NE.	− 4	24	24	9	53
NE.— E.	− 1	23	26	7	55
E.— SE.	0	22	33	11	66
SE.-- S.	2	23	32	18	75
Shifts	− 5	206	235	97	633
Leaps	35	25	40	43	103

* *Einiges über das Dove'sche Drehungsgesetz* (Something about Dove's Law of Gyration). Pogg. Ann. lxviii. p. 417.

SPRING

	1730—1737	1737—1743	1749—1758	1760—1769	Sum
S.— SW.	12	3	13	16	44
SW.— W.	11	3	13	17	44
W.— NW.	9	8	19	12	48
NW.— N.	11	5	15	9	40
N.— NE.	12	9	17	15	53
NE.— E.	13	10	18	19	60
E.— SE.	13	4	14	23	54
SE.— S.	12	5	17	24	58
Shifts	93	47	126	135	401
Leaps	41	28	60	67	196

SUMMER

	1730—1737	1737—1743	1749—1758	1760—1769	Sum
S.— SW.	11	18	15	15	59
SW.— W.	10	18	16	15	59
W.— NW.	8	11	21	18	58
NW.— N.	8	13	22	21	64
N.— NE.	8	9	19	23	59
NE.— E.	8	9	16	20	53
E.— SE.	11	16	12	17	56
SE.— S.	9	15	10	17	51
Shifts	73	109	131	146	459
Leaps	33	26	51	46	156

AUTUMN

	1730—1737	1737—1743	1749—1758	1760—1769	Sum
S.— SW.	8	14	19	14	43
SW.— W.	8	12	22	23	65
W.— NW.	15	18	21	17	71
NW.— N.	18	17	21	18	74
N.— NE.	17	19	15	19	70
NE.— E.	17	19	11	13	60
E.— SE.	10	14	7	16	47
SE.— S.	9	15	8	13	47
Shifts	102	128	122	133	485
Leaps	27	22	36	30	115

The first column of the following table contains the figures for five other years, viz., 1729, 1744, 1747, an unknown year, and 1759; the second column those for the whole thirty-nine years.

	Five years	Thirty-nine years
S.—SW.	73	322
SW.—W.	63	318
W.—NW.	55	312
NW.—N.	50	300
N.—NE.	54	299
NE.—E.	51	292
E.—SE.	58	295
SE.—S.	65	109
Gyrations .	469	2347
Leaps . .	$80 + x$	$650 + x$

The last column gives us an annual average of $60\frac{1}{6}$ as the excess of the direct over the retrograde rotations; and if we take the number of leaps in the year 1829, in which they were not observed, at 19, an annual average of $17\frac{1}{6}$ leaps. Hence, if we count the leaps as equal to four times one shift in an undetermined direction, we have, as excess of the direct rotations, $60 \pm 58\frac{2}{3}$.

In addition to this, the number of direct shifts at the different seasons is nearly constant; perhaps it is a little higher in winter. As to the rate at which the shift takes place, it is evident that, as the wind must make an equal number of shifts through all points of the compass, it is inversely proportional to the frequency with which the wind blows from each point. If we assume $\frac{2}{3}$ as the number of leaps in the direction of direct gyration, and therefore 66·8 for that of the positive shifts, we see that, as the eight points are distant 45° from each other, the wind will veer $\frac{66·8}{8}$ times through each quarter, and in n times $\frac{337·5}{n}$ degrees. This gives, on an average —

S.	to SW.	. .	4·1°	N.	to NE.	. . 6·0°
SW.	W.	. .	3·1	NE.	E.	. . . 5·85 ·
W.	NW.	. .	3·6	E.	S.E.	. . 6·5
NW.	N.	. . .	4·55	SE.	S.	. . . 6·2

Liverpool. FOLLET OSLER[*] in the ' Report of the 25th Meeting of the British Association, held at Glasgow in September 1855,' has given the results which have been obtained under the direction of Hartnup at the Liverpool Observatory, by means of Osler's self-registering anemometer, improved by Robinson, in the years 1852—1855. The following complete revolutions were observed :—

Year	Direct	Retrograde	Excess of direct
1852	28	12	16
1853	24	12	12
1854	26	2	24
1855	24	10	14
Mean . .	25·5	9	16·5

Greenwich. The calculations are taken from Glaisher's ' Greenwich Magnetical and Meteorological Observations,' for eighteen years.

At Greenwich the average annual excess of the direct over the retrograde rotations for the first seventeen years, is fourteen complete revolutions. In the several years, 1842 to 1858, it was 13·0, 20·7, 21·6, 7·5, 18·1, 10·7, 12·1, 23·3, 15·9, 19·1, 8·8, —1·8, 6·8, 10·6, 16·3, 14·7, 24·0. Thus we see that the year 1853 exhibits a complete anomaly. As regards its temperature it presented such an exception to the general rule, that in eastern Germany the month of March was colder than that of February, which again was colder than January. In Berlin the temperature of the last-named two months was the same. It is

---

[*] An account of the self-registering Anemometer and Rain-gauge erected at the Liverpool Observatory in the autumn of 1851, with a summary of the records for the years 1852, 1853, 1854, and 1855.

interesting to find that such an unusual distribution of temperature is accompanied by abnormal motions of the vane to an extent which has not been observed in any other year since the introduction of self-registering instruments. In any calculations of averages for a short period, the year 1853 must be totally excluded.

In the following table I have added together the rotations expressed in degrees for the different years, and thence determined the absolute mean. We see that the great anomalies of the last years reduce the excess in autumn materially,— in fact remove it completely in October. This shows probably that cyclones are most common at this period.

(The year 1859 has been included in this table, by Professor Dove's desire. It does not appear in the German edition.)

	1842	1843	1844	1845	1846	1847	1848
January . . .	135	270	112	− 90	45	225	0
February . .	540	495	−506	855	− 720	−1,012	− 135
March	293	1,327	1,012	382	675	562	922
April	337	551	731	− 270	1,147	1,080	− 315
May	360	1,923	652	315	540	135	1,215
June	1,525	1,102	405	1,462	− 1,327	382	720
July	472	810	2,677	360	− 135	1,642	990
August . . .	270	1,260	607	− 180	1,012	− 22	1,282
September . .	161	90	2,589	225	292	202	180
October . . .	90	67	−112	−247	− 382	157	− 45
November . .	135	− 427	− 90	1,125	67	742	−652
December . .	382	− 22	−292	...	− 562	225	180

	1849	1850	1851	1852	1853	1854
January . . .	945	652	585	360	− 67	1,035
February . . .	360	157	900	360	− 247	292
March . . .	472	1,957	1,080	1,192	− 67	2,227
April	−472	922	742	360	− 1,057	382
May	910	67	1,957	−180	877	− 1,597
June	2,319	1,237	− 337	22	− 495	− 202
July	607	135	832	1,102	− 720	2,475
August . . .	1,282	315	427	292	1,035	1,710
September . .	675	− 90	652	45	337	− 2,970
October . . .	1,170	− 22	− 360	− 1,102	− 1,080	− 585
November . .	562	157	− 202	495	720	− 315
December . .	−427	225	607	225	90	0

	1855	1856	1857	1858	1859	Mean for 18 years
January. . . .	−225	405	−270	135	381·5	257
February . . .	−540	180	− 67	−202·5	22·5	41
March	180	810	1,035	1,237·5	495	868
April	−742	−720	495	2,137·5	401·5	239
May	1,192	180	1,125	−360	1,192·5	583
June	1,170	2,250	1,260	3,735	1,260	855
July	720	1,170	720	−585	1,305	810
August	923	495	0	472·5	22·5	622
September . . .	1,957	697	270	405	292·5	332
October. . . .	−225	427	45	135	67·5	−40
November . . .	−810	45	697	922·5	450	306
December . . .	270	− 90	− 22	652·5	−854·5	36

This gives annually fourteen complete revolutions.

Oxford (Meteorological Observations made at the Radcliffe Observatory). According to Johnson's observations in the year 1856, between the middle of July and the end of December, the ratio which the direct rotations bore to the retrograde was 24 : 19. In the year 1857 the excess of the former over the latter was 9 complete revolutions.

Brussels. QUETELET has calculated the years 1842 to 1846 for Brussels. He gives the following table of the excess of direct over retrograde rotations :—

	Winter	Spring	Summer	Autumn	Year
1842	2	5	12	2	21
1843	1	0	8	1	11
1844	0	7	2	1	10
1845	0	4	5	1	10
1846	1	8	8	−1	16
Mean	0·8	4·8	7·0	1	14

The shortest period of a complete revolution was thirty-nine hours, the longest eighty-eight days; the ratio of the direct to the retrograde changes in the several months was — 0·97, 1·00, 1·06, 2·89, 1·47, 2·00, 2·45, 2·18, 1·53, 1·30, 0·75, 1·58. January and November are accordingly abnormal, especially the latter.

Kharkov. LAPSHINE, in a special treatise ' *Les Vents qui souflent à Kharkov suivent-ils la loi decouverte par M. Dove?'* (Do the winds which blow at Kharkov obey the law discovered by M. Dove or not?) finds the excess of rotations for the mean of five years (1845—1849) to be fifteen complete revolutions, and for the separate months the ratios of the excesses to be as follows: 25·2, 5·6, 19·8, 36·2, 32·8, 35·8, 16·8, 21·4, 14·2, 16·2, 12·4, 4·5. February and December minimum, April, May, and June maximum, and the year 1846 abnormal.

The number of complete revolutions at Liverpool, London, Brussels, and Kharkov being 16·5, 13·6, 14, 15 respectively, we see that there is a correspondence between the results obtained in western and eastern Europe ; indeed the fact could hardly be otherwise, inasmuch as the phenomenon is produced by the alternation of polar and equatorial currents of great lateral extension. This lateral extension of the current is also proved by the results of my investigations on the non-periodical changes of temperature, from which it appears that when too high temperatures are felt temporarily over any area, they are always compensated by too low temperatures over some adjoining district, but that a very extensive area of observation is always required in order to pass beyond the boundaries of one of the currents.

Madrid. RICO Y SINOBAS* gives the following number of complete revolutions obtained from the Osler's anemometer in 1854.

	Direct	Retrograde	Excess
Winter . . .	12	6	6
Spring . . .	11	8	3
Summer . .	26	5	21
Autumn . .	16	5	11
Year . . .	65	24	41

* *Resumen de los Trabajos Meteorologicos correspondientes al anno* 1854, *verificados en el Real Observatorio de Madrid* (Summary of Meteorological Observations during the year 1854, verified at the Royal Observatory of Madrid), 1854, 4to.

Bermuda.[*] Excess of direct shiftings over indirect, in degrees

	1859	1860	Mean
January	...	1,710	...
February . . .	...	1.035	...
March	...	1,732	...
April	540	2,002	1,271
May	945	1,147	1,046
June	495	315	405
July	—90	1,057	483
August	450	967	709
September . . .	585	675	630
October	1,800	...	...
November . . .	2,155	...	...
December . . .	1,260	...	...
	38 complete revolutions.		

From the investigations as to the storms on the W. coast
of Europe which will follow, we may conclude that the
majority of these storms follow on the whole the course of
the storm of December 24, 1821, which I proved in the
year 1828 to have been a cyclone, and which we shall
consider in full at p. 162. They move from SW. to NE.,
and frequently parts of England lie on the W. side of the
path of the centre. The phenomenon of two winds 'fighting,' which we shall examine after the cyclones, appears,
according to all observations which I have as yet been
able to obtain, to be more common in central and eastern
Europe. The storms of the Mediterranean and Black Seas
appear to possess preeminently the character of a contest
between the return Trade-wind on its descent from the
higher regions of the atmosphere, with the polar current
which moves in the opposite direction.

This leads us to the conclusion that the chief cause of
concealment of the law of gyration in eastern Europe is

[*] Anemometry at Bermuda (Eighth Number of Meteorological Papers
published by authority of the Board of Trade).

the 'fighting' of currents; in western Europe it is cyclonical disturbance.

In conclusion I subjoin the results, expressed in degrees,
which have been obtained from the Osler's anemometer at
Bombay.   The normal phenomenon makes its appearance
here especially at the change of the Monsoon, but is observable at all seasons.   Although Hadley's principle may
be applied to the change of the Monsoon, yet we cannot
apply it immediately to the lesser oscillations of the currents once they have set in ; consequently the question in
how far these are due to local causes is left to the observer
to determine.

	1848	1849	1850	1851	Mean
January . . .	709	0	720	720	537
February . .	1,080	720	1,080	337	643
March . . .	1,766	1.440	720	1,103	1,257
April	1,091	1,080	−45	1,035	790
May . . .	315	382	−68	23	163
June	372	270	697	−450	222
July	709	1,080	360	765	729
August . . .	382	180	1,080	360	500
September . .	383	1,238	472	1,125	805
October . . .	2,169	1,462	3,228	1,463	2,093
November . .	1,800	945	720	360	706
December . .	1,080	923	720	45	692

The final result of the foregoing investigation proves
clearly that the law of gyration, any disturbing influences
notwithstanding, is noticeable in a decided manner in the
direct observations of the vane.

I may be excused for having quoted such a large number of proofs of the regular gyration of the wind, when it
is remembered that my object has been to establish a fact
which, when I published in 1827 my first investigations
on the Law of Gyration, was distinctly denied by Schouw,
and which too was not mentioned in the meteorological

works of Cotte, Deluc, Saussure, Dalton, Daniell, or Howard. Ten years subsequently Pouillet expressed his opinion on the subject in the following words:—' *On a cru remarquer que dans certains lieux les vents se succèdent dans un ordre déterminé; mais ces observations présentent encore trop d'incertitude pour qu'il nous soit permis de les discuter ici.*' (Some observers believe that at certain stations the winds succeed each other in a determined order; however, these observations exhibit as yet too great uncertainty for us to discuss them here.)

3. THE REGULAR SHIFTING OF THE WIND IN THE SOUTHERN HEMISPHERE.

$\rightarrow$ S., SE., E., NE., N., NW., W., SW., S. $\rightarrow$

1. I am indebted to the kindness of Captain WENDT, who sailed round the world several times, as commander of the Prussian ship *Prinzess Luise*, in answer to an enquiry addressed to him, for the following notice:—

'The wind in the southern hemisphere usually veers from N. through W. to S. and SE. Its direction consequently is contrary to that of the wind in the northern hemisphere. To the best of my knowledge the fact is nearly as follows:— Near the Cape of Good Hope in summer, the wind is chiefly SE., but if the wind turns northerly it always freshens. When the best summer months are at an end, after a calm of short duration the wind usually blows very moderately from SE., with an unusually clear sky. As soon as the wind veers towards the E. it rises steadily; and if it has got to the N., clouds and lightning are sure to appear on the western horizon, and in less than half an hour a storm from WNW. will almost always ensue, and will not cease until, after 24 or 48 hours, it has veered more to the S.

'Near Cape Horn, both to the eastward and westward of it, with a north wind, there is generally fine weather; when it veers

to the NW. it soon blows hard; with a WNW. to SW. it usually
blows a storm (which is also frequently the case from WNW.
with a NW. wind succeeding it). The wind subsides as it becomes
southerly; SSE. brings fine weather, frequently succeeded by
a calm.'

2. LE GENTIL.* *Ethiopian Sea.* ' On the 25th and 26th, we
experienced a squall, which veered from N. through W. to SW.,
and I remarked a fact which you have had opportunity of observing more frequently than myself, that the winds do not follow the
same rule in this hemisphere as in the northern; in the latter they
make the circuit of the compass from N. to NE., E., SE., S., &c.;
in the southern hemisphere, on the contrary, they move in an
opposite direction. The thunderstorms, gales, and squalls seem
to me to bear out the analogy in both hemispheres. Physicists
have as yet given no explanation of this phenomenon.'

3. DON ULLOA.† *Pacific Ocean.* 'The wind in the South
Pacific Ocean is never fixed in the NE., nor does it ever change
from thence to the E.; its constant variation being to the W. or
SW., contrary to what is observed in the northern hemisphere.
In both the change of wind usually corresponds with the course
of the sun; hence, in the northern, it changes from E. to S. and
thence to W., in the southern from E. to N. and thence to W.'

4. FORSTER.‡ *South Sea.* 'Between 40° and 60° S. latitude,
in the year 1773, we quite unexpectedly met with easterly winds,
which were very contrary to our course at the time. It was also
remarkable that every time the wind changed, which was the
case four times between the 5th of June and the 5th of July, it
gradually moved half round the compass in a direction contrary
to the course of the sun.' I believe I may understand Forster
to have borrowed this expression, in the way usual among sailors,
from the course of the sun in the northern hemisphere.

5. DON COSME CHURRUCA.§ *Straits of Magellan.* (Commu-

---

* *Voyage dans les Mers de l'Inde* (Voyage in Indian Waters), vol. ii. p.
701, Letter to M. de la Nux.

† Voyage to South America, vol. i. p. 8, ch. 3.

‡ *Bemerkungen* (Remarks), p. 111.

§ *Apendice á la Relacion del Viage al Magelhanes* (Appendix to the Story
of the Voyage to Magellan). Madrid, 1793, p. 15.

nicated by Baron von Humboldt.) 'In the southern hemisphere the winds usually succeed each other in the opposite order to that which they follow in ours. In our seas the winds change from N. to E., S., W., to N.; in the southern hemisphere, on the contrary, from N. to W., S., E., to N.'

6. HORSBURGH.* *South Atlantic Ocean*, Lat. 38°. 'Although here the westerly winds prevail during most months of the year, they are often very unsettled, completing a revolution round the horizon, coincident with the course of the sun, every two, three, or four days with intervening calms, particularly when the wind is from the south-west quarter. When cloudy weather accompanies these northerly and north-westerly winds, there is a risk of a sudden shift to south-west and south.'

7. *Sea South of the L'Agulhas Bank.*† Around the Cape Bank, as in the open sea far to the SW., SE., and southward of the Cape, the winds in changing follow the course of the sun, seldom veering from north to eastward, &c., but mostly from NW. to W., SW., and southward. After blowing strong from NW. or W., if the wind should veer to SW. and southward it becomes light or is succeeded by a calm. If a light breeze continues, it veers to the south-eastward, where it may keep fixed for a considerable time, but not above a day most probably if it is the winter season. From south-east it veers to the east and NE., then to NNE. and north.

* * * * * * *

When the wind at SE. or ESE. shifted to the north-eastward, the Dutch commanders were directed by the Company to take in the mainsail. If lightning appeared in the NW. quarter, they were to wear and shorten sail; for in the first case they expected a hard gale at north-west, and if lightning was seen in that direction they thought the gale would commence in a sudden shift or whirlwind, which might be fatal if they were taken aback.'

8. *Sea South of Australia.*‡ 'Off the south coast of Terra

---

* East India Directory, vol. i. p. 67.
† Ibid. vol. i. p. 91.
‡ Ibid. vol. i. p. 97.

Australis the progress of the gales is usually this: the barometer
falls to 29·5 ins. or lower, and the wind rises from the NW. with
thick weather, commonly with rain; it veers gradually to the
west, increasing in strength, and when it veers to the southward
of that point, the weather begins to clear up. At south-west
the gale blows hardest and the barometer rises, and by the time
the wind gets to S. or SSE. it becomes moderate, with fine
weather and the barometer above 30 inches.'

9. KING & FITZROY.* *South Coast of Chili.* 'With northerly
and north-westerly winds the sky is overcast, the weather
unsettled, damp, and disagreeable.   These winds are always
accompanied by clouds and usually by thick rainy weather.
From the north-west the wind in general shifts to the south-west
and thence to the southward.   Sometimes it flies round in a
violent squall, accompanied by rain, thunder, and lightning. At
other times it draws gradually round.   Directly the wind is southward of west the clouds begin to disperse, and as a steady southerly
wind approaches, the sky becomes clear and the weather healthily
pleasant.   A turn of fresh southerly wind is usually followed by
a moderate breeze from the south-west, with very fine weather.
Light variable breezes follow, clouds gradually overspread the
sky, and another round turn is generally begun by light or
moderate north-easterly breezes, with cloudy weather and often
rain.

'This is the general order of change; when the wind shifts
against this order, or backs round, bad weather with strong winds
may be expected.'

10. FITZROY.† *Terra del Fuego.* 'From the north the wind always begins to blow moderately, but with thicker weather and more
clouds than when from the eastward; it is generally accompanied
by small rain.   Increasing in strength, it draws to the westward
gradually, and blows hard from between north and north-west with
heavy clouds, thick weather, and much rain.   When the fury
of the north-wester is expended, which varies from twelve to

* Narrative of the Surveying Voyages of the Adventure and Beagle,
Appendix to vol. ii. p. 184.

† Ibid. p. 314. Sailing Directions for Terra del Fuego.

fifty hours, or even while it is blowing hard, the wind sometimes shifts suddenly into the SW. quarter, blowing harder than before. This wind soon drives away the clouds, and in a few hours causes clear weather, though perhaps with heavy squalls passing occasionally. All kinds of shifts and changes are experienced from north to south, by the west, during the summer months. The barometer is lowest with NW. winds and highest with SE. It is low with NW. and N.; if it fall to 29 in., or 28·8, a SW. gale may be expected, but it does not commence until the column has ceased to descend.'

11. *Sea between Cape Horn and* 40° *S. Lat.*[*] 'If the sky becomes overcast during a calm, which is usually of short duration, the first breeze rises from N. and NNE., it dies gradually away, rain begins to fall, and the weather becomes thick, especially near the coast. If the wind has changed to NW., it generally veers to WSW., often with heavy showers, or a shower follows soon, and the wind reaches its greatest height. Whenever these showers from WSW. have lasted for a time, they cease with a SW. wind, and the weather becomes fine; they then change, but seldom, to SSW. and SSE. The last-mentioned changes take place chiefly near the coast, and to the south-west of Cape Horn.'

12. DUPETIT THOUARS.[†] *Valparaiso.* 'In winter the winds are changeable; if they come from NE. and N. they are accompanied by rain and clouds, if they freshen from the N. they veer to the NW. with showers, thence to the W. and S., which brings back the fine weather.'

The observations made in the roadstead of Valparaiso lead to the same result as those of Dupetit Thouars: *i.e.*—If after a calm a breeze rises from N. and NNE., it veers round to NW., and then to SW., S., and SE.

13. HEYWOOD.[‡] *Rio de la Plata.* 'Before a SW. gale or a ".Pampero," the weather is usually uncertain, with changeable winds from the N. and NW. quarters, preceded by a

* *Dépôt Général de la Marine. Instructions sur les Côtes du Pérou en* 1824 (Instructions on the Coasts of Peru in 1824), p. 7.
† .*Plan de la Baie de Valparaiso* (Plan of the Bay of Valparaiso).
‡ Instructions and Observations for Navigating the Rio de la Plata.

considerable fall of the barometer.   This rises a little, however, before the wind goes to SW., and often continues to rise even when the wind blows strongly from this quarter.'

14. BASIL HALL.* 'To express myself generally I will remark, that I have often noticed, that in the southern hemisphere the wind changes more frequently from S. to E., N., W., S., than in the opposite direction, while in the northern hemisphere it changes more frequently from S. to W., N., E., S.'

15. DUMONT D'URVILLE. 'Extract from the Log of the *Astrolabe* relating to the principal variations of the wind in the southern hemisphere, in the years 1826—1827' (private letter), Toulon, August 3, 1837, at the moment of starting on his last voyage round the world.

1826.   'From the 10th to the 13th of August (Lat. S. 30°, Long. 23° W.).   The wind which was WSW., and light, veered to S. and SSE., where it grew stronger, and then to ESE. and NE., where it was light again.'

1826.   '14th to 16th August (Lat. S. 31°, Long. 16° W.).   Wind first light from NE. and NW., very strong from WNW. to SSW., and then more moderate from S. and SSE.'

1826.   '19th to 30th August (Lat. S. 33°—37°, Long. 13° W.— 29° E.).   The wind blows with great violence from NW., W., and SW., a heavy gale from NW., which moderates and veers on the next day to SW., SSW., S., SSE., and NE.'

1826.   '6th to 11th September (Lat. S. 37°—38°, Long. 50° E.). Strong NE. and NNE., with fine weather, then N., NW., and W.'

1826.   '8th to 24th October (Lat. S. 39°, Long. 115° E.).   Wind in general high from NW. and SW.   On one occasion from the 16th to the 19th it veered gradually through the whole compass from NW. to SW., S., SE., SSE., NE., NNW., NW.'

1826.   '(Retrograde) (Lat. S. 38°, Long. 122° E.).   A very violent wind from NE., shifts suddenly round to SSE., and changes on the following day to SSW. and W., where it dies away.'

* From a private letter.

1826. '5th November (Lat. S. 39°, Long. 135° E.). A NNE. wind changes to NNW. and moderates, veering on the next day through SSW. to SSE. and E.'

1826. '19th to 28th November (Lat. S. 39°, Long. 142°—148° E.). The winds were not strong, but they veered three times from right to left through the whole compass, *i.e.* from N. through W. to S., and from S. through E. to N.'

1826. '29th November to 2nd December (Lat. S. 39°, Long. 148° E.). The wind veers again twice in the same direction.'

1827. '5th to 9th January (Lat. S. 40°—43°, Long. 160° E.). The wind at first strong from NE., blows afterwards violently from NW. and WNW., rises to a gale from S. and SSE., and then moderates.'

1827. '12th to 16th February (Lat. S. 35°, Long. 176° E). A WNW. and W. wind veers to S., then E. and NE., where it increases to a hurricane; afterwards it veers to NW. again and turns to W., and dies away when it gets to SW.'

1827. '13th March. Bay of Islands (Lat. S. 35°, Long. 171° E.). The wind, which was high from NNW. and NW., goes round to WSW. and SW., and dies away when it passes from S. to SE. and SSE.'

1827. '(Retrograde) 31st March (Lat. S. 33°, Long. 177° E.). A violent N. wind changes to NE., E., SE., SSW., and SW., continues strong at that point, and at last dies away there.'

M. Dumont d' Urville adds to these extracts from his journal the following remarks:—'M. Dove may see that out of 18 well-marked cases only 2* appear to be in opposition to the law of change from N. to S. by W., and from S. to N. by E. I can recall to my memory very well, that whenever we had high winds from NW. or SW. we expected to see them moderate when they approached the south. M. Dove may be assured that, on

---

* These were probably cyclones.

my next voyage, I shall cause the officers of the Astrolabe to
observe accurately the direction in which the wind changes in
the southern hemisphere, and it is possible that I may address a
note on that subject from Valparaiso to M. von Humboldt, who
will communicate it to M. Dove.'

16. LEICHHARDT. *Australia* (private letter). Sydney, June 18,
1842. 'In Sydney I thought of your theory for the wind, and
found, as I expected, that it holds for this place in the opposite
direction. The hot NW. winds in Sydney are invariably followed by a violent S. or SW. wind, which brings the loose sand
from the hills opposite Botany Bay in dense clouds of dust to
the town. This wind is called 'a brick-fielder,' since it blows
from the brick-fields to the town. It is cold, and the change
of temperature produced by it is sometimes extraordinary,
$50° — 60°$ F. Mr. Clarke finds that all *squalls* change from
S. to W., N., and E., while the *regular* change of wind is from
S. to E., N., and W.'

17. STRELECKI,[*] *Australia and Van Diemen's Land,* remarks :
'The meteorological journals of Port Macquarie, Port Jackson,
Port Philip, and Port Arthur, show equally that, at each of
the stations, the winds, when they change their direction, do
this always in one fixed direction, viz., from left to right of the
meridian, when the face is turned to the equator. These observations confirm the law propounded by Dove, in his "*Meteorologische Untersuchungen,*" not only as regards the rotation in
the opposite directions on the two hemispheres, but also the
converse effect on the barometer, thermometer, hygrometer,
and rain-gauge.'

18. HORNER.[†] *Montevideo.* 'The wind changes in general
from E. to N., thence to W. and S. The N. and NW. winds are
the hottest, the S. and SE. the coldest.'

19. BYRON DRURY.[‡] *New Zealand.* 'The changes of the
wind take place almost invariably with the sun, or in the

[*] Physical Description of New South Wales and Van Diemen's Land,
p. 165.

[†] Medical Topography of Brazil and Uruguay. Philadelphia, 1845, p. 20.

[‡] On the Meteorology of New Zealand. First number of Meteorological
Papers published by authority of the Board of Trade, 1857, p. 65.

opposite direction to the motion of the hands of a watch; they are consequently opposite to those on the northern hemisphere, though they shift there, too, with the sun.'

20. VON WÜLLERSDORFF-URBAIN.* *Southern Hemisphere.* 'At certain intervals we had a SE. wind, but after a longer or shorter period it shifted through E., NE., and N. towards NW., W., SW., and sometimes even to S. Such a shift as this occupied several days. The wind never exceeded ordinary intensity excepting at St. Paul and near the Cape of Good Hope.'

21. *Melbourne.†* Excess of direct shiftings over retrograde (in complete revolutions). The direction of the rotation is reversed.

1858		1858		1859	
March . .	9·2	August . . .	1·8	January . .	4·8
April . . .	4·2	September . .	6·2	February . .	2·8
May . . .	2·1	October . .	1·5		
June . . .	2·6	November . .	3·3		
July . . .	3·5	December . .	6·2		

22. VAN GOGH.‡ *Southern Hemisphere.* 'When the wind is south-easterly and high, and is accompanied by drizzling rain, the barometer rises slowly to the height of about 30·25 inches, when we may expect that it will shift to NE. (at least, I have as yet only once observed that the wind, after passing through SE., shifted back towards SW.). When the wind has got to NE. it remains at that point, blowing with the same violence, and the weather is the same as before. The barometer now begins to fall in the same proportion as it had risen before, until it comes to the height of about 29·75 inches, then the wind

---

* *Beitrag zur Theorie der Luftströmungen* (Contribution to the Theory of Atmospheric Currents), 1858, p. 11.

† Results of the Magnetical, Natural, and Meteorological Observations made at the Flagstaff Observatory, Melbourne.

‡ *Uitkomsten van wetenschap en ervaring aangaande winden en zeestroomingen in sommige gedeelten van den Oceaan, uitgegeven door het Kon. Meteorologischen Instituut* (Results of Science and Experience regarding Winds and Marine Currents in certain parts of the Ocean, published by the Royal Meteorological Institute), Utrecht, 1857, p. 50.

shifts through N., and blows hardest when it has got to NNW.
The barometer now remains steady, and this last wind continues
for a few hours.   It subsequently becomes westerly, while the
barometer rises slowly.   Sometimes the wind gets to the south of
west and blows very hard, while the barometer rises rapidly.'

If we recapitulate our authorities, we find that we have
direct numerical determinations made by Schouw at
Apenrade, Eisenlohr at Carlsruhe, Emsmann at Berlin,
Kölbing at Gnadenfeld, Hofmeister at Lenzburg, Buys
Ballot in Holland; and anemometrical results from Follet
Osler at Liverpool, Glaisher at Greenwich, Johnson at
Oxford, Quetelet at Brussels, Lapshine at Kharkov, Rico y
Sinobas at Madrid, and from the observations at Bermuda
and at Melbourne.   In addition to these we shall give the
indirect proofs, from the changes of the meteorological
instruments, which are still more conclusive than the above,
and we have confirmatory evidence for both hemispheres
from the following sources : —

NORTHERN HEMISPHERE.	SOUTHERN HEMISPHERE.
Aristotle — Greece.	Don Ulloa — Pacific Ocean.
Theophrastus — Greece.	Le Gentil —Indian Ocean.
Pliny — Italy.	Forster — South Sea.
Bacon — England.	Don Cosme Churruca — S. Sea near S. America.
Mariotte — France.	Horsburgh — S. Atlantic Ocean.
Sturm — Germany.	———— — Sea S. of Australia.
Toaldo — Italy.	Capt Wendt — The Cape and Cape Horn.
Poitevin — Montpelier.	King and Fitzroy — S. Coast of Chili.
Kant — E. Prussia.	Fitzroy — Terra del Fuego.
Romme — N. Atlantic Ocean.	*Dépôt Général de la Marine* — Cape Horn — 40° S. Lat.
Lampadius —Freiberg.	Dupetit Thouars — Bay of Valparaiso.

NORTHERN HEMISPHERE.

SOUTHERN HEMISPHERE.

Dove —Königsberg.

Heywood — Rio de la Plata.

Schübler — Germany.

Basil Hall — General.

Dumont d'Urville — General.

Hildreth — N. America.

Leichhardt — Australia.

Duden — State of Missouri.

Strelecki— Van Diemen's Land.

Von Wrangel — Sitka.

Horner — Montevideo.

Dupuits de Maconet — S. of France.

Byron Drury — New Zealand.

Drake — N. America.

Kane — Smith Sound.

Von Wüllersdorff-Urbain — Southern Hemisphere.

Lartigue — Both hemispheres in his 'Theory of the Wind.'

Van Gogh — Southern Hemisphere.

This shows us that the old sailors' saying holds good on both hemispheres, viz : —

> When the wind veers against the sun,
> Trust it not, for back it will run.

Since the regular rotation of the wind is as clearly observable in England (Bacon) as in the interior of Russia (Lapshine); in Sitka, on the W. coast of N. America (Von Wrangel) as in the valley between the Rocky Mountains and the Alleghanies (Drake); in the southern hemisphere, on the coast of Chili (Fitzroy) as at the mouth of the Rio de la Plata (Horner); and is reported by the most experienced navigators as quite unmistakable on all the great oceans, it must be an appearance which is universal, independent of the relative position of the oceans and continents, and also independent of the mean direction of the wind and its annual variations. The effect on the consequent variations of the meteorological instruments changes at the different seasons, since, as I proved in the year 1827, for Paris, the coldest point of the windrose in Europe is nearer the NE. in winter, and nearer the NW.

in summer; and accordingly the warmest point in winter
is nearer SW., in summer nearer SE.   However, in order
to retain the general character of my nomenclature, I call
one of the alternating currents Polar, the other Equatorial.
The determination of the extremes of the windrose which
has been chosen for the annual mean variations of the
instruments, viz., NE. and SW. for the northern, SE. and
NW. for the southern hemisphere, must therefore be
modified for the different seasons and months, according
to the relative position of the station in question on the
western or eastern coast of a continent.

4.— VARIATIONS OF THE METEOROLOGICAL INSTRUMENTS
DEPENDENT ON THE LAW OF GYRATION.

IF we calculate the mean of all the observations of the
barometer, thermometer, and hygrometer, corresponding
to each of the different winds, after elimination of their
periodical variations,— i. e., if we determine the mean distribution of pressure, temperature, and saturation in the
windrose,— or, in other words, construct a barometrical,
thermal, and hygrometrical windrose, we shall find that
it has two poles of pressure and temperature.   There are
in the windrose two points, nearly diametrically opposite
to each other, at one of which the thermometer is lowest
and the barometer highest, and at the other the thermometer is highest and the barometer lowest.   The
mean heights of the barometer and thermometer corresponding to the different winds (*die barom. und therm.*

Windmittel) decrease uninterruptedly from the maximum
to the minimum of pressure, and from the maximum to
the minimum of temperature.   The first point is situated
near NE., the second near SW.   If we proceed from SW.
through W. to NE., the mean height of the barometer
increases, while that of the thermometer decreases ; if we
proceed further from NE. through E. to SW., the mean
height of the thermometer increases, while that of the
barometer decreases.*   These facts, which are observed
in the mean heights of the barometer and thermometer,
corresponding to the various winds, must also be traceable in the transitions of the same one into the other, *i.e.*
in the mean variations of the barometer and thermometer,
as well under the hypothesis of a changeable velocity of
rotation as of one which is constant.   Inasmuch as the
tension of aqueous vapour with regard to its distribution
in the windrose is closely related to the thermal, and the
pressure of the dry air to the barometrical windrose, it
follows that the changes of pressure of the dry air and
of the barometer are exactly related, in inverse proportion, to the changes of the temperature of the air and of
the tension of aqueous vapour therein contained.   If we
assume as a necessary consequence of the foregoing theoretical considerations, that the NW. wind plays the same
part on the southern hemisphere as the SW. wind here —

---

* In the northern hemisphere such windroses have been calculated for
several stations, viz. : —

Barometrical for Irkutsk, Archangel, St. Petersburg, Moscow, Stockholm,
Arys, Dantzic, Bütsow, Zechen, Berlin, Halle, Mühlhausen, Carlsruhe,
Middelburg, Utrecht, Hamburg, Apenrade, Copenhagen, Paris, London,
Chiswick, Oxford, Dublin, and Reikiavik.

Thermal for Yakutsk, Irkutsk, Tara, Moscow, Archangel, St. Petersburg,
Stockholm, Arys, Conitz, Zechen, Hamburg, Halle, Mühlhausen, Buda,
Carlsruhe, Utrecht, Paris, London, Chiswick, Oxford, Dublin, and Sitka.

Hygrometrical for Arys, Halle, London, Chiswick, Oxford, and Dublin.

that the SE. wind there corresponds to the NE. wind here,* — we obtain the following laws : —

MEAN VARIATIONS OF THE METEOROLOGICAL INSTRUMENTS.

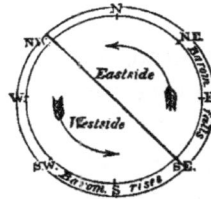

NORTHERN HEMISPHERE.

1. The barometer falls with E., SE., and S. winds; with a SW. wind it ceases to fall and begins to rise; it rises with W., NW., and N. winds; and with a NE. wind it ceases to rise and begins to fall. (Fig.1.)

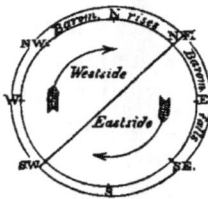

SOUTHERN HEMISPHERE.

1. The barometer falls with E., NE., and N. winds; with a NW. wind it ceases to fall and begins to rise; it rises with W., SW., and S. winds; and with a SE. wind it ceases to rise and begins to fall. (Fig. 2.)

NORTHERN HEMISPHERE.

FIG. 1.

SOUTHERN HEMISPHERE.

FIG. 2.

NORTHERN HEMISPHERE.

FIG. 3.

SOUTHERN HEMISPHERE.

FIG. 4.

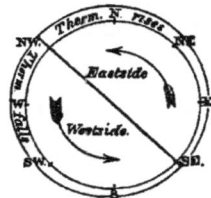

* This assumption is confirmed by the Dutch observations, calculated by Eijsbroek, and published in the *Ondersoekingen met den Zee Thermometer* (Experiments with the Marine Thermometer), 1861, p. 108. The following

NORTHERN HEMISPHERE.

2. The thermometer rises with E., SE., and S. winds; with a SW. wind it ceases to rise and begins to fall; it falls with W., NW., and N. winds; and with a NE. wind it ceases to fall and begins to rise. (Fig. 3.)

3. The tension of aqueous vapour increases with E., SE., and S. winds; with a SW. wind it ceases to increase and begins to decrease; it decreases with W., NW., and .N. winds; and with a NE. wind it ceases to decrease and begins to increase. (Fig. 3.)

4. The pressure of dry air decreases with E., SE., and S.

SOUTHERN HEMISPHERE.

2. The thermometer rises with E., NE., and N. winds; with a NW. wind it ceases to rise and begins to fall; it falls with W., SW., and S. winds; and with a SE. wind it ceases to fall and begins to rise. (Fig. 4).

3. The tension of aqueous vapour increases with E., NE., and N. winds; with a NW. wind it ceases to increase and begins to decrease; it decreases with W., SW., and S. winds; and with a SE. wind it ceases to increase and begins to decrease. (Fig. 4.)

4. The pressure of dry air decreases with E., NE., and N.

are the mean heights of the barometer in inches, corresponding to the different directions of the wind : —

Latitude S.	30°—40°	40°—45°
N.	30·008	29·913
NNE.	30·047	29·956
NE.	30·047	29·996
ENE.	30·087	30·039
E.	30·114	30·055
ESE.	30·134	30·067
SE.	30·142	30·043
SSE.	30·130	30·024
S.	30·083	30·012
SSW.	30·055	29·953
SW.	30·004	29·930
WSW.	29·973	29·894
W.	29·961	29·906
WNW.	29·961	29·874
NW.	29·977	29·882
NNW.	30·000	29·882
Mean . . .	30·024	29·915

winds; with a SW. wind it ceases to decrease and begins to increase; it increases with W., NW., and N. winds; and with a NE. wind it ceases to increase and begins to decrease. (Fig. 1. p. 120.)

winds; with a NW. wind it ceases to decrease and begins to increase; it increases with W., SW., and S. winds; and with a SE. wind it ceases to increase and begins to decrease. (Fig. 2. p. 120.)

We see from the above rules that the *agreement* between the two hemispheres consists in the fact that the behaviour of the meteorological instruments is the same in the two hemispheres for easterly, and equally so for westerly winds. The *difference* between the two hemispheres is only quantitative in the case of NW., NE., SW., and SE. winds, while it is qualitative in the case of N. and S. winds. This implies that the mean variations of the meteorological instruments, in the northern hemisphere, have a maximum value with NW. and SE. winds, and a minimum value with NE. and SW. winds (owing to a mutual compensation of opposite fluctuations); and, in the southern hemisphere, have a maximum value with NE. and SW. winds, and a minimum value with NW. and SE. winds (owing to a compensation of opposite fluctuations): whereas the mean variations with N. and S. winds in the northern hemisphere differ, according to their sign, from those with N. and S. winds in the southern hemisphere, while they agree, according to their magnitude, in both hemispheres, under similar conditions of climate. Accordingly, if an instrument rises with a N. wind in the northern hemisphere, it will fall with a N. wind in the southern, and *vice versâ*. The same holds good for S. winds.

The proofs for the laws above stated are contained in the subjoined tables.

1. THE VARIATIONS OF THE BAROMETER.

a. *The Northern Hemisphere.*

LAW. — The barometer falls with E., SE., and S. winds ; with a SW. wind it ceases to fall and begins to rise ; it rises with W., NW., and N. winds ; and with a NE. wind it ceases to rise and begins to fall. (Fig. 1. p. 120.)

In the following tables the sign $(+)$ indicates a rise, $(-)$ a fall. The sources for the different data are as follows : —

PARIS (calculated by me) from Dove, '*Meteorologische Untersuchungen über den Wind*' (Meteorological Investigations relative to the Wind). Poggendorff's Annalen, 1827, vol. xi. p. 535.

LONDON (calculated by me) from Dove, '*Ueber die von der Windesrichtung abhängigen Veränderungen der Dampfatmosphäre*' (On the Variations of Atmospherical Vapour depending on the direction of the Wind). Pogg. Ann. xvi. p. 288.

CHISWICK (calculated by me) from Dove, '*Ueber die Allgemeine Theorie des Windes*' (On the General Theory of the Winds). Reports of the Academy of Berlin, 1857, p. 90.

HALLE (calculated by Kaemtz) from Kaemtz, '*Vorlesungen über Meteorologie*' (Lectures on Meteorology), p. 311.

ZECHEN (calculated by Gube) from the Reports of the Academy of Berlin, 1857, p. 296.

DANTZIC (calculated by Galle) from Galle, '*Zur Prüfung des von Dove aufgestellten Gesetzes über das verschiedene Verhalten der Ost- und Westseite der Windrose*' (A Test of the Law laid down by Dove on the Differences observable between the East and West Side of the Windrose). Pogg. Ann. xxxi. p. 465.

ARYS (calculated by Vogt) from the Reports of the Academy of Berlin, 1857, p. 90.

ST. PETERSBURG (calculated by Kaemtz) from a private letter.

DORPAT (calculated by Kaemtz) from a private letter.

OGDENSBURGH (calculated by Coffin) from Meteorological Observations made at Ogdensburgh in the 'Returns of Meteorological Observations made for the year 1838 by sundry Academies in the State of New York.'

TORONTO (calculated by Dörgens) from the third vol. of the 'Observations, &c., made at Toronto.'

	PARIS		LONDON	CHISWICK	HALLE	ZECHEM	DANTZIC	ARYS	ST. PETERSBURG
Years	5	10	3	16		3	15	11	2½
Fluctuation	In 12 hours		Morning to Evening		In 16 hours	16	12	8	14
	Inches	Inches	Inches	Inches	Inches	Inches	Inches	Inches	Inches
WSW.	+ 0·0014	+ 0·0026	+ 0·011	+ 0·045	+ 0·0258	+ 0·0391	+ 0·0139	+ 0·0124	+ 0·0300
W.	+ 0·0425	+ 0·0393	+ 0·032	+ 0·064	+ 0·0719	+ 0·0826	+ 0·0052	+ 0·0489	+ 0·0626
WNW.	+ 0·0460	+ 0·0536	+ 0·049	+ 0·078	+ 0·0320	+ 0·0489	+ 0·0429	+ 0·0346	+ 0·0970
NW.	+ 0·0479	+ 0·0456	+ 0·018	+ 0·021	+ 0·0027	− 0·0195	+ 0·0436	+ 0·0222	+ 0·0800
NNW.	+ 0·0436	+ 0·0540	− 0·012	+ 0·005	− 0·0169	− 0·0595	+ 0·0589	− 0·0071	− 0·0205
N.	+ 0·0187	+ 0·0116	− 0·049	− 0·055	− 0·0453	− 0·0648	+ 0·0333	− 0·0258	− 0·0415
NNE.	− 0·0045	− 0·0064	− 0·048	− 0·076	− 0·0338	− 0·0568	+ 0·0068	− 0·0355	− 0·0536
NE.	− 0·0056	− 0·0092	− 0·023	− 0·017	− 0·0036	− 0·0169	+ 0·0276	− 0·0240	− 0·0536
ENE.	− 0·0311	− 0·0458					− 0·0086		
E.	− 0·0430	− 0·0501					− 0·0069		
ESE.	− 0·0512	− 0·0549					− 0·0020		
SE.	− 0·0476	− 0·0461					− 0·0109		
SSE.	− 0·0273	− 0·0456					− 0·0343		
S.	− 0·0396	− 0·0447					− 0·0457		
SSW.	− 0·0458	− 0·0445					− 0·0444		
SW.	+ 0·0047	− 0·0107					− 0·0078		

DORPAT.   In a communication entitled 'The Barometrical Windrose at Dorpat' (a letter to Prof. Dove), Kaemtz has calculated the results of the observations carried on by Mädler and himself at that station from Sept. 1, 1842, to Dec. 1859, in the following manner.   In the case of each direction of the wind, he has calculated the mean height of the barometer on the day on which the wind was observed, and also on the two days which succeeded, and the two days which preceded the observation.   If n be the number of the days (which is taken as positive for those which succeeded, negative for those which preceded the observation), he has calculated eight equations, of the form $a + b\,n + c\,n^2 + d\,n^3$ for each month, in which the co-efficients $a\ b\ c\ d$ represent the constants deduced from the observations.   Kaemtz has given the values, which are thus obtained, as differences from the mean values for the month, and has given those for all the observations, and for the wet days and dry days, separately.   In the following table I have myself deduced the fluctuations in order to make the results comparable with those which are given above.   In the table headed 'Day before,' the number indicates the fluctuation of the barometer, expressed in inches, during the twenty-four hours which preceded the observation, the sign $(+)$ expressing a rise, $(-)$ a fall.   In the table headed 'Day after,' the number indicates in a similar way the fluctuation of the barometer during the twenty-four hours which succeeded the observation.   To the mean result for the year I have added the values obtained for the days next but one before and after the observation.   The observations for the rain have been calculated in the same manner.

MEAN OF ALL OBSERVATIONS.

DAY BEFORE.

	NW.	N.	NE.	E.	SE.	S.	SW.	W.
January .	+ 0·106	+ 0·179	+ 0·134	+ 0·122	− 0·046	− 0·003	− 0·098	− 0·024
February .	+ 0·082	+ 0·150	+ 0·119	+ 0·046	− 0·027	− 0·090	− 0·106	− 0·035
March . .	+ 0·042	+ 0·103	+ 0·106	+ 0·057	− 0·018	− 0·080	− 0·091	− 0·041
April . .	− 0·001	+ 0·052	+ 0·090	+ 0·052	− 0·021	− 0·070	− 0·069	− 0·033
May . . .	− 0·002	+ 0·014	+ 0·054	+ 0·032	− 0·039	− 0·066	− 0·030	− 0·008
June . . .	+ 0·007	+ 0·002	+ 0·022	+ 0·006	− 0·050	− 0·066	− 0·018	+ 0·043
July . . .	+ 0·038	+ 0·019	+ 0·008	− 0·008	− 0·046	− 0·066	− 0·027	+ 0·027
August . .	+ 0·059	+ 0·052	+ 0·026	− 0·003	− 0·039	− 0·066	− 0·042	+ 0·019
September .	+ 0·065	+ 0·096	+ 0·070	+ 0·011	− 0·036	− 0·062	− 0·053	− 0·003
October . .	+ 0·082	+ 0·126	+ 0·098	+ 0·035	− 0·031	− 0·076	− 0·072	− 0·006
November .	+ 0·081	+ 0·170	+ 0·142	+ 0·026	− 0·058	− 0·073	− 0·065	− 0·022
December .	+ 0·100	+ 0·186	+ 0·145	+ 0·026	− 0·060	− 0·086	− 0·080	− 0·020
Year . . .	+ 0·053	+ 0·097	+ 0·085	+ 0·026	− 0·041	− 0·073	− 0·060	− 0·011

DAY AFTER.

	NW.	N.	NE.	E.	SE.	S.	SW.	W.
January .	+ 0·088	+ 0·209	+ 0·068	+ 0·003	− 0·036	− 0·154	+ 0·019	− 0·015
February .	+ 0·079	+ 0·101	+ 0·059	+ 0·020	− 0·011	− 0·064	− 0·085	− 0·014
March . .	+ 0·052	+ 0·077	+ 0·056	+ 0·027	− 0·006	− 0·053	− 0·071	− 0·020
April . .	+ 0·016	+ 0·047	+ 0·041	+ 0·015	− 0·022	− 0·037	− 0·028	− 0·017
May . . .	0·000	+ 0·015	+ 0·027	− 0·010	− 0·046	− 0·027	− 0·004	0·000
June . . .	− 0·002	− 0·004	− 0·003	− 0·032	− 0·029	− 0·028	+ 0·017	+ 0·067
July . . .	+ 0·026	− 0·001	− 0·021	− 0·037	− 0·050	− 0·037	+ 0·002	+ 0·033
August . .	+ 0·051	+ 0·029	− 0·006	− 0·027	− 0·041	− 0·045	− 0·018	+ 0·029
September .	+ 0·063	+ 0·072	+ 0·034	− 0·015	− 0·043	− 0·046	− 0·027	+ 0·015
October . .	+ 0·077	+ 0·098	+ 0·060	+ 0·001	− 0·041	− 0·057	− 0·039	+ 0·014
November .	+ 0·066	+ 0·130	+ 0·094	− 0·012	− 0·067	− 0·048	− 0·025	− 0·003
December .	+ 0·077	+ 0·132	+ 0·086	− 0·010	− 0·060	− 0·056	− 0·042	− 0·003
Year . . .	+ 0·047	+ 0·068	+ 0·043	− 0·007	− 0·042	− 0·043	− 0·031	+ 0·003

YEAR.

	BEFORE		AFTER	
Days	2	1	1	2
NW.	− 0·017	+ 0·053	+ 0·047	− 0·035
N.	+ 0·034	+ 0·097	+ 0·068	− 0·051
NE.	+ 0·062	+ 0·085	+ 0·042	− 0·065
E.	+ 0·044	+ 0·026	− 0·007	− 0·053
SE.	+ 0·010	− 0·041	− 0·042	+ 0·010
S.	− 0·013	− 0·073	− 0·046	+ 0·067
SW.	− 0·032	− 0·060	− 0·031	+ 0·058
W.	− 0·040	− 0·011	+ 0·003	+ 0·003

IN TIME OF RAIN.

DAY BEFORE.

	NW.	N.	NE.	E.	SE.	S.	SW.	W.
January .	+ 0·041	+ 0·170	+ 0·123	− 0·008	− 0·101	− 0·138	− 0·143	− 0·075
February .	+ 0·060	+ 0·165	+ 0·121	+ 0·005	− 0·073	− 0·120	− 0·150	− 0·091
March . .	+ 0·129	+ 0·126	+ 0·087	− 0·001	+ 0·014	− 0·084	− 0·121	− 0·084
April . .	− 0·019	+ 0·064	+ 0·034	− 0·023	− 0·046	− 0·052	− 0·067	− 0·051
May . . .	+ 0·002	+ 0·011	+ 0·006	− 0·041	− 0·055	− 0·043	− 0·030	− 0·008
June . . .	+ 0·010	− 0·012	− 0·019	− 0·034	− 0·061	− 0·056	− 0·003	+ 0·022
July . . .	+ 0·026	− 0·002	− 0·003	− 0·008	− 0·051	− 0·075	− 0·032	+ 0·025
August . .	+ 0·041	+ 0·028	+ 0·023	− 0·024	− 0·036	− 0·088	− 0·062	+ 0·009
September .	+ 0·050	+ 0·060	+ 0·043	+ 0·017	− 0·035	− 0·090	− 0·082	− 0·010
October . .	+ 0·055	+ 0·088	+ 0·059	− 0·003	− 0·059	− 0·095	− 0·087	− 0·025
November .	+ 0·059	+ 0·046	+ 0·027	− 0·023	− 0·093	− 0·111	− 0·094	− 0·036
December .	+ 0·066	+ 0·148	+ 0·102	− 0·023	− 0·112	− 0·131	− 0·116	− 0·053
Year . . .	+ 0·041	+ 0·081	+ 0·053	− 0·060	− 0·065	− 0·090	− 0·082	− 0·032

DAY AFTER.

	NW.	N.	NE.	E.	SE.	S.	SW.	W.
January .	+ 0·100	+ 0·192	+ 0·116	− 0·007	− 0·041	− 0·041	− 0·073	− 0·041
February .	+ 0·103	+ 0·288	+ 0·126	+ 0·010	− 0·027	− 0·042	− 0·082	− 0·046
March . .	+ 0·091	+ 0·164	+ 0·101	+ 0·004	− 0·080	− 0·028	− 0·059	− 0·028
April . .	+ 0·066	+ 0·106	+ 0·055	− 0·019	− 0·029	− 0·010	− 0·013	+ 0·001
May . . .	+ 0·037	+ 0·051	+ 0·020	− 0·034	− 0·036	− 0·002	+ 0·025	+ 0·027
June . . .	+ 0·027	+ 0·026	+ 0·014	− 0·024	− 0·046	− 0·012	+ 0·027	+ 0·040
July . . .	+ 0·035	+ 0·030	+ 0·033	− 0·003	− 0·031	− 0·035	+ 0·011	+ 0·042
August . .	+ 0·051	+ 0·047	+ 0·051	+ 0·027	− 0·028	− 0·054	− 0·014	+ 0·038
September .	+ 0·069	+ 0·065	+ 0·053	+ 0·025	− 0·027	− 0·057	− 0·025	+ 0·036
October . .	+ 0·079	+ 0·086	+ 0·045	− 0·002	− 0·034	− 0·045	− 0·022	+ 0·028
November .	+ 0·083	+ 0·114	+ 0·054	− 0·026	− 0·048	− 0·035	− 0·026	+ 0·010
December .	+ 0·051	+ 0·156	+ 0·082	− 0·027	− 0·050	− 0·035	− 0·045	− 0·018
Year . . .	+ 0·069	+ 0·102	+ 0·063	− 0·006	− 0·036	− 0·033	− 0·024	+ 0·007

YEAR.

	BEFORE		AFTER	
Days	2	1	1	2
NW.	− 0·017	+ 0·041	+ 0·069	− 0·026
N.	+ 0·034	+ 0·081	+ 0·102	− 0·051
NE.	+ 0·062	+ 0·053	+ 0·063	− 0·065
E.	+ 0·044	− 0·010	+ 0·006	− 0·053
SE.	+ 0·010	− 0·065	− 0·036	+ 0·010
S.	− 0·013	− 0·090	− 0·033	+ 0·012
SW.	− 0·032	− 0·082	− 0·024	+ 0·058
W.	− 0·041	− 0·032	+ 0·007	+ 0·002

It will be seen that in both of the above tables there is a complete confirmation of the laws laid down by me in the year 1827. At the same time we see that if the observations be extended to a period of five days, the change of wind which then takes place passes from the rising to the falling side of the windrose and *vice versâ*. In the summer months, when the precipitation of rain, owing to the *courant ascendant* is more common, it is natural to expect that the rules for the occurrence of rain during the several currents will present greater simplicity, and that the change in the distribution of barometrical pressure in the windrose, owing to the alteration in position of the isothermals at the different seasons, will exhibit itself clearly. Both these facts are observable in the tables.

Mr. Dörgens has made calculations extending over a longer period of time (1846–48), which are published in the 3rd volume of 'The Observations made at the Magnetical and Meteorological Observatory at Toronto in

OGDENSBURGH — STATE OF NEW YORK.

Fluctuation	Barometer	Thermometer	Fluctuation	Barometer	Thermometer
	In 1 hour			In 1 hour	
	Inches	Degrees F.		Inches	Degrees F.
SW.	+ 0·0006	− 0·018	NE.	− 0·0044	− 0·015
SW. by W.	+ 0·0025	− 0·055	NE. by E.	− 0·0058	+ 0·094
WSW.	+ 0·0027	− 0·018	ENE.	− 0·0077	+ 0·115
W. by S.	+ 0·0057	− 0·081	E. by N.	− 0·0072	+ 0·077
W.	+ 0·0052	− 0·063	E.	−· 0·0062	+ 0·103
W. by N.	+ 0·0065	− 0·069	E. by S.	− 0·0061	+ 0·162
WNW.	+ 0·0091	− 0·252	ESE.	− 0·0051	+ 0·146
NW. by W.	+ 0·0104	− 0·281	SE. by E.	− 0·0040	+ 0·114
NW.	+ 0·0111	− 0·322	SE.	− 0·0051	+ 0·140
NW. by N.	+ 0·0091	− 0·306	SE. by S.	− 0·0065	+ 0·145
NNW.	+ 0·0080	− 0·276	SSE.	− 0·0065	+ 0·138
N. by W.	+ 0·0080	− 0·236	S. by E.	− 0·0060	+ 0·161
N.	+ 0·0033	− 0·197	S.	−· 0·0074	+ 0·314
N. by E.	+ 0·0039	− 0·165	S. by W.	− 0·0055	+ 0·177
NNE.	+ 0·0007	− 0·144	SSW.	− 0·0036	+ 0·162
NE. by N.	− 0·0017	− 0·063	SW. by S.	− 0·0014	+ 0·065

TORONTO.

		Barometer			Thermometer			Tension of Aqueous Vapour		
		9—12	12—3	9—3	6—12	12—6	6—6	8—12	12—4	8—4
NE.	11	−0·036	−0·027	−0·063	−4·99	+4·14	−0·85	−0·014	+0·015	+0·001
NE. by E.	7	+0·008	−0·013	−0·005	−2·15	−0·49	−2·64	−0·009	−0·008	−0·017
ENE.	18	−0·006	−0·028	−0·034	−5·28	+0·42	−4·86	−0·021	0·001	−0·021
E. by N.	21	−0·013	−0·028	−0·041	−2·95	+0·75	−2·20	−0·008	−0·009	−0·017
E.	52	−0·024	−0·026	−0·050	−0·53	+0·28	−0·25	−0·002	+0·005	+0·003
E. by S.	25	−0·020	−0·028	−0·048	+0·39	+0·23	+0·62	+0·007	−0·003	+0·004
ESE.	35	−0·015	−0·018	−0·033	+3·10	−0·23	+2·87	+0·003	−0·010	−0·007
SE. by E.	12	−0·018	−0·031	−0·049	+3·57	−1·35	+2·22	+0·014	−0·015	−0·001
SE.	14	−0·006	−0·025	−0·031	+3·71	−0·93	+2·78	+0·019	+0·004	+0·023
SE. by S.	17	−0·021	−0·032	−0·053	+3·12	−0·29	+2·83	+0·023	−0·001	+0·022
SSE.	29	−0·025	−0·018	−0·043	+4·22	−0·72	+3·85	+0·029	−0·009	+0·020
S. by E.	25	−0·007	−0·026	−0·033	+6·77	+0·87	+5·90	+0·014	+0·006	+0·020
S.	28	−0·011	−0·020	−0·031	+5·08	−1·21	+3·87	+0·029	−0·006	+0·023
S. by W.	34	−0·011	−0·016	−0·027	+4·19	−1·16	+3·03	+0·030	−0·023	+0·007
SSW.	57	−0·012	−0·018	−0·030	+3·19	−0·96	+2·23	+0·016	−0·015	+0·001
SW. by S.	21	−0·036	−0·024	−0·060	+1·21	0·69	+1·90	+0·011	+0·013	+0·024
SW.	34	−0·016	−0·012	−0·028	+0·23	−0·56	−0·33	+0·001	−0·001	+0·002
SW. by W.	8	−0·017	+0·005	+0·012	−4·21	0·58	−3·63	−0·007	−0·004	−0·001
WSW.	24	−0·002	+0·015	+0·013	−1·82	0·62	−1·20	−0·014	+0·005	−0·009
W. by S.	18	+0·009	+0·029	+0·038	−0·27	−3·12	−3·39	−0·025	−0·010	−0·035
W.	23	+0·038	+0·045	+0·083	−2·25	−1·52	−3·77	−0·025	+0·006	−0·019
W. by N.	13	+0·022	+0·035	+0·057	1·01	−1·52	−0·51	−0·032	−0·008	−0·040
WNW.	25	+0·041	+0·040	+0·081	0·42	−2·84	−2·42	−0·028	+0·009	−0·019
NW. by W.	16	+0·041	+0·053	+0·094	−1·78	−3·64	−5·42	−0·037	−0·025	−0·062
NW.	30	+0·051	+0·049	+0·100	−4·10	−1·40	−5·50	−0·040	−0·002	−0·042
NW. by N.	16	+0·031	+0·035	+0·066	−3·15	+0·60	−2·55	−0·024	+0·018	−0·006
NNW.	29	+0·027	+0·017	+0·044	−4·52	+0·11	−4·41	−0·035	+0·003	−0·032
N. by W.	24	+0·016	+0·024	+0·040	−3·04	+0·25	−2·79	−0·024	+0·016	−0·008
N.	26	+0·018	+0·017	+0·035	−4·24	−0·31	−4·55	−0·032	+0·007	−0·025
N. by E.	12	+0·004	+0·020	+0·024	−4·44	+0·13	−4·31	−0·024	+0·007	−0·017
NNE.	5	+0·036	+0·027	+0·063	−6·94	−0·83	−7·77	−0·017	−0·025	−0·042
NE. by N.	3	−0·005	+0·027	+0·022	−5·86	+1·22	−4·64	−0·026	+0·025	−0·001

Canada.' These observations commenced in the year 1840, and it would be very desirable that the calculation should be extended to the whole series of years. The fluctuations of the barometer, thermometer, and hygrometer have been determined for the periods given at the head of the table, after elimination of the diurnal variation. The period is rather short, considering that the directions are referred to the whole thirty-two points of the compass ; but yet the Law of Gyration is clearly traceable.

K

It will now be easy to trace the same laws in the imperfect observations of former meteorologists.

GUNZENHAUSEN, *Luz* *, says : 'The north and north-west winds cause the barometer to rise, I may almost say invariably. The east and north-east winds frequently produce the same effect, but not so invariably as the preceding. These winds are accompanied by a clear sky. The barometer also rises with a west wind, and then the sky is often covered with scattered clouds at a great height, from which rain rarely falls. With a south-east wind the barometer falls, and yet the weather does not change as long as the wind does not shift towards the south. No such certain rules can be given for south and south-west winds. The barometer usually falls when the wind is in this quarter; but if it has been long in the same point, and especially if there has been wet weather for some time, the barometer rises again, even though the wind continues to blow from the south and south-west. In the same way I have observed the barometer to fall sometimes with a north or east wind, when the wind had been for a long time in the same point, and the clear weather was about to break and give way to clouds and rain.'

Holland. This subject has been examined by VAN SWINDEN † more completely than by Luz. Horsley ‡ had been the first to demonstrate in a more definite manner the influence of the direction of the wind upon the state of the barometer, which had been already pointed out by Halley and Mariotte, by calculation of a barometrical windrose. The attention of Van Swinden being excited by this, he proposed to himself the question : 'How often does the barometer rise with a certain wind, and how often does it fall with the same?' The results of his calculation are a necessary consequence of the law of gyration. He finds in the year 1779 that the barometer

---

* *Beschreibung von Barometern* (Description of Barometers), 1784, p. 351.

† *Mémoires sur les Observations Météorologiques faites à Franecker en Frize pendant* 1779 (Memoirs on the Meteorological Observations made at Franecker in Friesland during the year 1779).

‡ An abridged State of the Weather at London in the year 1774.— *Philosoph. Transact.* for 1775.

	Rose		Fell	
With SW.	74	times . . .	83·9	times
W.	36	,, . . .	16·6	,,
NW.	83	,, . . .	43·5	,,
N.	12	,, . . .	9·3	,,
NE.	24	·, . . .	28·0	,,
E.	1	·· . . .	8·3	,,
SE.	18	,, ...	51·8	,,
S.	10	,, . . .	15·5	,,

In the three preceding years he had obtained with regard to
W., NW., N., E., SE., and S., results coinciding with the above;
but, on the contrary, deviations with regard to NE. and SW.
These turning-points are therefore here as clearly remarkable
as in Luz's observations. It does not appear, from any observation of Van Swinden's, that he was himself acquainted with
the law of gyration.

Had Van Swinden, instead of asking, ' How often does the
barometer rise with a given wind?' asked instead, ' Does the
fall exceed the rise, or the converse, when a given wind blows?'
a clear course of reasoning must have led him to the explanation
of the phenomenon. Even Saussure could not account for it;
for he asks in his Hygrometrie, ' Why do E. winds, although
cold and dry, cause the barometer to fall in England and
Holland, according to the observations of Horsley[*] and Van
Swinden, while W. winds cause it to rise? I am acquainted
with no hypothesis which explains this fact satisfactorily.'

It is probable that the annual mean direction of the
wind at the place of observation has no effect on the
result, from the fact that while in London this direction
is due W., in Paris and Dantzic it is WSW. It is, however, more important to decide whether the result is not
effected by the change in the mean direction of the wind
at the different seasons. It appears from the following
tables that the law of fluctuation of the barometer is
independent of the yearly periods.

---

[*] With regard to Horsley Saussure is in error: he had only calculated
the mean, not the rise and fall.

PARIS.

	NE.	E.	SE.	S.	SW.	W.	NW.	N.	Correction
January	+0·010	−0·053	−0·043	−0·044	−0·025	+0·055	+0·112	+0·027	0·157
February	−0·014	+0·025	−0·050	−0·002	−0·044	+0·041	+0·059	−0·005	0·023
March	−0·005	−0·033	−0·077	−0·069	−0·040	+0·042	+0·113	+0·038	0·015
April	−0·013	−0·046	−0·051	−0·011	+0·026	+0·034	+0·066	+0·011	0·019
May	+0·010	−0·104	−0·051	−0·039	+0·018	+0·019	+0·013	−0·019	0·014
June	−0·019	−0·050	−0·059	−0·075	+0·002	+0·028	+0·003	−0·001	0·014
July	−0·037	−0·084	−0·043	−0·044	+0·001	+0·030	+0·058	+0·006	0·015
August	−0·009	−0·028	−0·031	−0·053	+0·022	+0·042	+0·020	+0·022	0·021
September	−0·018	−0·060	−0·067	−0·014	−0·010	+0·068	+0·013	+0·021	0·014
October	−0·002	−0·050	−0·015	−0·017	+0·007	+0·017	+0·037	+0·065	0·009
November	+0·049	−0·106	−0·041	−0·175	−0·012	+0·031	+0·065	+0·082	0·008
December	−0·010	−0·007	−0·039	−0·092	+0·005	+0·071	+0·079	+0·048	0·009
Year	−0·009	−0·050	−0·046	−0·045	−0·008	+0·039	+0·046	+0·011	0·011

The observations at Dantzic give the following fluctuations for 16 winds :—

	NNE.	NE.	ENE.	E.	ESE.	SE.	SSE.	S.
January	+0·030	+0·032	...	−0·010	−0·021	+0·027	−0·053	−0·027
February	+0·109	+0·178	+0·053	+0·012	−0·002	+0·007	−0·031	−0·049
March	+0·008	+0·089	−0·031	+0·041	−0·008	−0·047	−0·004	−0·038
April	+0·026	+0·083	−0·033	−0·059	−0·008	−0·047	−0·004	−0·056
May	−0·002	−0·020	−0·020	+0·007	−0·017	−0·013	−0·059	−0·063
June	−0·041	+0·015	−0·019	−0·024	−0·030	−0·021	−0·024	−0·028
July	+0·012	−0·035	+0·012	−0·035	+0·016	−0·064	0·000	−0·055
August	+0·012	+0·039	−0·001	+0·014	−0·065	+0·006	−0·103	−0·053
September	−0·011	+0·019	0·000	−0·008	+0·006	−0·042	+0·003	−0·036
October	+0·018	−0·004	−0·093	+0·028	+0·037	−0·004	−0·045	−0·073
November	−0·293	+0·185	−0·066	−0·093	−0·068	−0·112	−0·011	−0·043
December	−0·038	−0·022	...	+0·110	+0·016	+0·013	−0·014	−0·040

	SSW.	SW.	WSW.	W.	WNW.	NW.	NNW.	N.
January	−0·015	−0·030	−0·006	+0·006	+0·007	+0·073	+0·071	+0·008
February	−0·084	−0·086	−0·103	−0·006	−0·062	+0·073	+0·037	+0·047
March	−0·150	−0·004	+0·002	+0·005	+0·045	+0·024	+0·175	+0·068
April	+0·027	−0·021	+0·016	−0·066	+0·017	+0·041	+0·111	+0·039
May	−0·065	+0·012	−0·046	−0·004	+0·076	+0·030	+0·046	+0·020
June	−0·013	−0·026	−0·065	+0·036	+0·017	+0·022	−0·060	+0·017
July	+0·009	+0·021	+0·049	+0·004	+0·060	+0·012	+0·034	+0·017
August	±0·000	−0·029	+0·012	±0·000	+0·015	+0·022	−0·009	+0·024
September	+0·003	+0·004	+0·018	+0·020	−0·001	+0·039	+0·040	+0·040
October	−0·073	+0·034	+0·102	+0·021	÷0·077	+0·050	+0·028	+0·050
November	−0·022	−0·017	+0·072	+0·027	+0·077	+0·076	+0·050	+0·046
December	−0·081	+0·004	+0·102	+0·046	+0·132	+0·075	+0·126	+0·118

CHISWICK.

	W.	NW.	N.	NE.	E.	SE.	S.	W.
January . .	+ 0·050	+ 0·029	+ 0·093	+ 0·021	− 0·013	− 0·029	− 0·029	− 0·067
February . .	+ 0·032	+ 0·051	+ 0·059	+ 0·047	+ 0·005	− 0·056	− 0·072	− 0·031
March . . .	+ 0·030	+ 0·027	+ 0·035	+ 0·025	+ 0·003	− 0·042	− 0·089	− 0·032
April . . .	+ 0·005	− 0·015	+ 0·038	− 0·007	+ 0·013	+ 0·003	− 0·028	− 0·082
May . . .	+ 0·025	− 0·027	+ 0·024	− 0·016	− 0·019	− 0·014	− 0·106	+ 0·003
June . . .	+ 0·022	+ 0·042	+ 0·006	+ 0·012	− 0·031	− 0·027	− 0·048	− 0·001
July	+ 0·042	+ 0·025	+ 0·032	+ 0·019	− 0·019	− 0·093	− 0·051	− 0·004
August . .	+ 0·013	+ 0·171	+ 0·071	− 0·003	− 0·003	− 0·030	− 0·078	+ 0·014
September. .	+ 0·032	+ 0·060	+ 0·103	− 0·002	− 0·002	− 0·057	− 0·065	+ 0·002
October . .	+ 0·021	+ 0·067	+ 0·099	+ 0·016	+ 0·026	− 0·081	− 0·014	− 0·020
November .	+ 0·079	+ 0·061	+ 0·024	+ 0·043	+ 0·043	− 0·008	− 0·080	− 0·023
December . .	+ 0·066	+ 0·088	+ 0·104	+ 0·040	+ 0·040	− 0·012	− 0·012	+ 0·015
Winter. . .	+ 0·049	+ 0·056	+ 0·085	+ 0·036	+ 0·001	− 0·032	− 0·038	− 0·028
Spring . . .	+ 0·018	− 0·005	+ 0·037	+ 0·001	− 0·001	− 0·011	− 0·074	− 0·012
Summer . .	+ 0·025	+ 0·079	+ 0·036	+ 0·009	− 0·005	− 0·050	− 0·059	+ 0·003
Autumn . .	+ 0·044	+ 0·063	+ 0·075	+ 0·019	+ 0·022	− 0·073	− 0·053	− 0·014

ARYS.

	W.	NW.	N.	NE.	E.	SE.	S.	SW.
January . .	+ 0·028	+ 0·131	+ 0·061	+ 0·037	+ 0·011	− 0·019	− 0·049	− 0·034
February . .	− 0·021	+ 0·096	+ 0·096	+ 0·097	− 0·025	− 0·041	− 0·053	− 0·021
March . . .	− 0·020	+ 0·060	+ 0·067	+ 0·038	− 0·002	− 0·019	− 0·047	− 0·058
April . . .	− 0·015	+ 0·028	+ 0·029	+ 0·017	− 0·011	− 0·020	− 0·028	− 0·017
May . . .	+ 0·019	+ 0·022	+ 0·019	+ 0·008	− 0·016	− 0·027	− 0·014	− 0·007
June . . .	+ 0·017	+ 0·023	+ 0·008	− 0·012	− 0·012	− 0·033	− 0·034	− 0·018
July . . .	+ 0·022	+ 0·025	+ 0·006	− 0·004	− 0·020	− 0·044	− 0·027	− 0·019
August . .	+ 0·025	+ 0·042	+ 0·032	+ 0·004	− 0·019	− 0·036	− 0·039	− 0·018
September .	+ 0·012	+ 0·033	+ 0·028	+ 0·004	− 0·006	− 0·030	− 0·039	− 0·015
October . .	+ 0·036	+ 0·066	+ 0·077	+ 0·074	− 0·020	− 0·037	− 0·031	− 0·001
November .	+ 0·026	+ 0·062	+ 0·081	− 0·005	− 0·005	− 0·014	− 0·016	− 0·032
December . .	+ 0·033	+ 0·061	+ 0·057	+ 0·053	+ 0·017	− 0·017	− 0·039	− 0·036
Winter. . .	+ 0·009	+ 0·093	+ 0·073	+ 0·061	+ 0·005	− 0·025	− 0·047	− 0·031
Spring . . .	− 0·009	+ 0·036	+ 0·037	+ 0·018	− 0·010	− 0·021	− 0·030	− 0·036
Summer . .	+ 0·021	+ 0·028	+ 0·013	− 0·004	− 0·017	− 0·037	− 0·034	− 0·018
Autumn . .	+ 0·024	+ 0·049	+ 0·052	+ 0·019	− 0·010	− 0·026	− 0·028	− 0·016
Year . . .	+ 0·012	+ 0·049	+ 0·035	+ 0·022	− 0·007	− 0·026	− 0·036	− 0·024

B.— *The Southern Hemisphere.*

LAW.— The barometer falls with E., NE., and N. winds; with a NW. wind it ceases to fall and begins to rise; it rises with W., SW., and S. winds; and with a SE. wind it ceases to rise and begins to fall. (Fig. 2, p. 120.)

M. Galle * has given a direct confirmation of this law by a calculation of observations recorded in the log-book of Captain Wendt's vessel, on her voyage round the globe, in the years 1830 to 1832.

	Southern Hemisphere	Northern Hemisphere	Southern Hemisphere
	Inches	Inches	Inches
WSW.	+ 0·001	+ 0·003	+ 0·004
W.	− 0·000	+ 0·008	− 0·001
WNW.	− 0·000	+ 0·013	− 0·004
NW.	− 0·001	+ 0·019	− 0·004
NNW.	− 0·001	+ 0·019	− 0·004
N.	− 0·001	+ 0·008	− 0·004
NNE.	− 0·001	− 0·008	− 0·004
NE.	− 0·002	− 0·008	− 0·002
ENE.	− 0·002	− 0·009	− 0·001
E.	− 0·001	− 0·007	− 0·000
ESE.	− 0·000	− 0·006	+ 0·000
SE.	+ 0·001	− 0·006	+ 0·001
SSE.	+ 0·002	− 0·007	+ 0·002
S.	+ 0·002	− 0·011	+ 0·005
SSW.	+ 0·002	− 0·011	+ 0·007

The observations of Jansen† agree with these; he says:—

In twenty-one cases it appears that, in general, when the wind shifted from NE. through N. towards NW., it was accompanied by a falling barometer; and that when, on the contrary,

---

* *Bestätigung der Dove'schen Windtheorie durch die Barometer-verän-derungen der Südlichen Halbkugel* (Confirmation of Dove's Theory of the Winds by the Fluctuations of the Barometer in the Southern Hemisphere).— Pogg. *Ann.* xxxviii. 476.

† *Ondersoekingen met den Zee Thermometer* (Investigations with the Marine Thermometer), 1861, p. 108.

it shifted from W. and NW. through N. towards NE., the barometer always rose. When a NE. wind became north-westerly it usually increased in violence, and the weather was squally; while when a NW. wind became north-easterly, the wind decreased, but the weather was still thick.

In thirty-two cases it appears that when the wind shifted from SW. towards S. and SE., and the barometer rose, the wind decreased in intensity; but when the SW. wind shifted through S. towards SE., with a falling barometer, it increased.

2. THE FLUCTUATIONS OF THE THERMOMETER.

A. — *Northern Hemisphere.*

LAW. — The thermometer rises with E., SE., and S. winds; with a SW. wind it ceases to rise and begins to fall; it falls with W., NW., and N. winds; and with a NE. wind it ceases to fall and begins to rise. (Fig. 3, p. 120.)

In the following tables the sign $(+)$ indicates a rise, $(-)$ a fall.

	PARIS	CHISWICK	HALLE	ZSCHEN
S.	$-0°65$	$+0°16$	$+0°97$	$+2°09$
SSW.	$-0·27$			
SW.	$-1·13$	$+0·16$	$-0·05$	$-0·59$
WSW.	$-1·40$			
W.	$-1·37$	$-0·45$	$-1·13$	$-1·42$
WNW.	$-2·32$			
NW.	$-0·23$	$-0·45$	$-1·06$	$-2·54$
NNW.	$-0·90$			
N.	$+0·20$	$-0·05$	$-0·25$	$-1·76$
NNE.	$+0·41$			
NE.	$+1·10$	$-0·29$	$+0·79$	$+0·41$
ENE.	$+1·67$			
E.	$+1·42$	$-0·07$	$+0·92$	$+1·37$
ESE.	$+4·25$			
SE.	$+2·48$	$+1·26$	$+1·28$	$+1·98$
SSE.	$+1·60$			

ZECHEN.

	Winter	Spring	Summer	Autumn
SW.	+ 0·59	− 0·38	− 1·85	− 0·65
W.	− 0·45	− 1·13	− 2·72	− 1·40
NW.	− 2·81	− 2·59	− 2·32	− 2·43
N.	− 2·18	− 1·80	− 1·13	− 1·76
NE.	− 0·63	+ 0·16	+ 0·90	+ 1·22
E.	+ 0·50	+ 1·37	+ 1·94	+ 1·67
SE.	+ 1·82	+ 2·75	+ 0·20	+ 3·15
S.	+ 3·02	+ 1·87	+ 1·15	+ 2·21

CHISWICK.

	Winter	Spring	Summer	Autumn
SW.	+ 0·99	+ 0·20	− 0·45	− 0·18
W.	− 0·77	+ 0·09	− 0·41	− 0·74
NW.	− 0·45	+ 0·47	+ 0·50	− 1·31
N.	− 0·59	− 0·41	+ 0·41	− 0·77
NE.	− 1·26	− 0·72	+ 0·83	− 0·25
E.	+ 0·11	− 0·47	+ 0·43	+ 0·45
SE.	+ 1·26	+ 0·83	+ 0·99	+ 2·00
S.	+ 0·63	+ 0·25	− 0·74	− 0·79

We see that at Paris, Halle, Zechen, and Ogdensburgh the law comes out very clearly, while at Chiswick this is only the case in winter. For this reason it is very desirable that a calculation should be made from some other journal of observations in England. In winter the coldest wind is from the NE., and in summer from the NW., and accordingly the point where the change from a rise to a fall (and vice versâ) takes place, undergoes a corresponding alteration.

The only proofs from the southern hemisphere are the indirect ones, especially those contained in the observations of Leichhardt and Strelecki, which have been before referred to. The only journal which has been published in detail is that of Hobarton in Van Diemen's Land ; but in it two opposite winds (NW. and SE.) appear to predominate to such an extent, that the values obtainable from it for the shiftings with each individual wind, would give too great a difference in the values to obtain an accurate mean. Nevertheless, it would be desirable to carry out the calculation. I have myself been unable to fulfil the promise of performing this task, which I gave at the Cambridge Meeting of the British Association in 1845 ; and I fear that there is no prospect of my being able to do so, owing to the small proportion of my time which I can devote to scientific investigations.

---

3. Variations in the Tension of Aqueous Vapour.

A.—Northern Hemisphere.

Law.—The tension of aqueous vapour increases with E., SE., and S. winds; with a SW. wind it ceases to increase and begins to decrease ; it decreases with W., NW., and N. winds; and with a NE. wind it ceases to decrease and begins to increase.

	London	Halle
	Inches	Inches
W.	0·000	−0·005
NW.	−0·007	−0·021
N.	−0·014	−0·016
NE.	−0·005	−0·010
E.	−0·010	+0·007
SE.	+0·006	+0·028
S.	+0·008	+0·016
SW.	+0·002	0·000

We have no journals kept in the southern hemisphere.

4. Variations in the Pressure of the dry Air.

A.—*Northern Hemisphere.*

Law.—The pressure of dry air decreases with E., SE., and S. winds; with a SW. wind it ceases to decrease and begins to increase; it increases with W., NW., and N. winds; and with a NE. wind it ceases to increase and begins to decrease.

	LONDON	HALLE
	Inches	Inches
W.	+ 0·011	+ 0·031
NW.	+ 0·039	+ 0·092
N.	+ 0·063	+ 0·048
NE.	+ 0·023	+ 0·013
E.	− 0·002	− 0·024
SE.	− 0·055	− 0·073
S.	− 0·050	− 0·050
SW.	− 0·025	− 0·004

There are no calculated journals for the southern hemisphere.

We may append to the above laws the following rules for the fall of rain or snow :—

(1.) The fraction of saturation of the air increases with easterly, decreases with westerly, winds, and this takes place rather beyond the points where the barometer and thermometer attain their extreme values. The ground of this is, that on the west side the cold wind, which sets in close to the ground, commences by diminishing the capacity of the air for moisture, while the warm wind on the east side, which sets in at a higher level, at once increases that capacity. Observations carried on at Paris for ten

years lead to the following results.   The sign (+) indicates a rise, (−) a fall.

NW.	−4˙5	SE.	+4˙7
N.	−7˙0	S.	+2˙7
NE.	−4˙3	SW.	+7˙6
E.	−0˙9	W.	+5˙0

(2.) The formation of the *cirrus*, the indication of the
commencement of the equatorial current which is visible
below, is accompanied by a fall of the barometer; that of
the *cumulostrati*, rolling up from the western horizon, by
a rise of the barometer.

For this statement the only proofs I can allege are the
meteorological observations made by me at Königsberg
from 1826 to 1829.   Since that time I have found that
the fact has been confirmed by my observations.

(3.) Inasmuch as the difference in temperature, and
consequently in pressure, as shown by the barometer,
between the currents is greatest in winter, the polar
current displaces the equatorial most rapidly at that
season.   During the precipitation which takes place at
this time, the lower current shifts more to the northward.
Hence the mean direction of the wind is more northerly
with snow than with rain.   In Berlin it is WSW. with
rain (65° 54′. S. being taken =0°), and NW. with snow
(133° 59′).

(4.) Owing to the same causes, which produce, in
general, a greater oscillation of the barometer in winter
than in summer, we find that on an average the barometer is more depressed below the mean level for the
wind during snow than during rain.   Leopold von Buch
was the first to observe this fact.   The results of the
observations at Berlin are—

	Mean	With Rain	With Snow
N.	29·869	29·700	29·597
NE.	29·896	29·761	29·640
E.	29·873	29·768	29·609
SE.	29·712	29·578	29·504
S.	29·580	29·495	29·376
SW.	29·629	29·536	29·479
W.	29·763	29·681	29·627
NW.	29·827	·29·755	29·696

If, however, both rain and snow fall during the course of the same circuit of the windrose, owing to the fact that the snow which falls when the wind is SE., changes to rain when it is SW., and returns to snow when the wind becomes westerly, we shall find the level of the barometer lower during the fall of rain than during that of snow. This fact proves the incorrectness of Espy's opinion, viz. that the reason that the level of the barometer is lower during snow than during rain is, that a greater amount of latent heat must be set free during the condensation of aqueous vapour to the form of snow, than when it is only condensed to rain.

(5.) The barometer falls with rain during easterly winds, rises with rain during westerly. I have found that the observations at Paris give the following results (in inches) for the fluctuation in twelve hours; (+) indicates a rise, (−) a fall.

NNE.	−0·0005	SE.	−0·0194	WSW.	+0·0008
NE.	+0·0148	SSE.	−0·0395	W.	+0·0942
ENE.	−0·0419	S.	−0·0276	WNW.	+0·1251
E.	−0·0561	SSW.	−0·0267	NW.	+0·1189
ESE.	−0·0302	SW.	−0·0065	NNW.	+0·0663
				N.	+0·0586

Kämtz has confirmed the results obtained at Paris by a calculation of the observations made at Stockholm.

	Day before rain	Day of rain
E.	− 0·001	− 0·036
SE.	− 0·044	− 0·058
S.	− 0·036	− 0·054
SW.	− 0·063	− 0·024
W.	+ 0·011	+ 0·019
NW.	+ 0·027	+ 0·094
N.	+ 0·037	+ 0·053
NE.	+ 0·005	+ 0·039

(6.) If we neglect the cold produced by evaporation, we find that the temperature rises after rain with an easterly wind, falls after rain with a westerly wind. The results of the observations are given below; the daily fluctuation is, of course, eliminated, as in all the previous cases.

NE.	+ 2·39°	SW.	− 2·12°
ENE.	+ 7·25	WSW..	− 1·22
E.	+ 0·02	W.	− 2·18
ESE.	+ 3·15	WNW.	− 2·16
SE.	+ 6·97	NW.	− 0·97
SSE.	+ 0·81	NNW..	+ 1·40
S.	+ 0·18	N.	+ 4·91
SSW.	+ 0·05	NNE.	− 1·31

The data which have been cited prove clearly that the contrast between the east and west sides of the windrose is as marked when a fall of rain or snow takes place as when the circuit of the wind is completed without such a precipitation of moisture.

We have referred the contrasts between the east and west sides of the windrose, which are so clearly marked as regards the precipitation of moisture, to the fact that the phenomena of the west side are produced by the displacement of the equatorial by the polar current, while those of the east side are due to the displacement of the polar by the equatorial current. The observations from

which the above rules have been deduced are affected, in
one respect, by a local influence,—viz., that Europe is
situated between the ocean on the west and a great continent on the east. This disturbing influence might be
eliminated by taking other stations whose position was an
exact contrast to that just described, and for this purpose
the coast stations of the United States are admirably
adapted. It is to be desired that even a single journal,
extending over a considerable number of years, should be
subjected to examination, so that the readings of the
barometer, thermometer, and hygrometer, taken a certain
number of hours before a definite observed direction of
the wind, which is accompanied by the precipitation of
moisture, should be compared with their readings taken
at the same number of hours after it. Such a calculation
would show us the effect which is produced when easterly
winds have increased their condition of saturation by their
passage over the sea, while the westerly winds have lost a
proportion of the aqueous vapour which they contained
by having been in contact with land, especially in a district where a north and south chain of mountains lies in
the path of the wind from the western coast. With the
exception of Toronto, which is not so well adapted for
such investigations, as it lies actually on the shore of a
great fresh-water lake, I am acquainted with no journal
from the district alluded to, excepting that published by
Mr. Bache in the Observations at the Meteorological
Observatory at Philadelphia. This journal does not extend over a sufficient period of time, inasmuch as we have
to deal with magnitudes of the second order, if we take
the northerly and southerly winds as magnitudes of the
first order. I said in the third volume of the Observations made at the Magnetical and Meteorological Observatory at Hobarton in Van Diemen's Land, p. 10 :—

Meteorology began with the study of the phenomena which are observed in Europe; it received its first extension of importance from the observations of the phenomena in tropical America (i. e. by Humboldt). If the facts which hold true for Europe were equally applicable to the temperate and frigid zones in all longitudes, and if, in like manner, tropical America represented the torrid zone in general, it would not be of much importance in what part of the world the study of meteorology took its origin. This is, however, not the case, and a too hasty generalisation has led to the disregard of some important phenomena, while those of less importance were brought into the foreground. It was time for this science to free itself from the leading-strings of its infancy.

It is to be hoped that the want which is here expressed may be supplied by cooperation between the land-stations on the different continents, especially since the naval observers have already set the example of united action. Such cooperation as this is the more necessary as there are many questions in meteorology which can only be answered by a complete and uninterrupted series of observations carried on at the same station.

We may conclude indirectly that the results which have been obtained for land hold good at sea, if the results of experience arrived at by seamen agree with those established by observations and calculations at land-stations. It was, therefore, satisfactory to me to find in a work published in the year 1855, by that distinguished navigator, Captain Lartigue,* which is a second edition of the work which appeared in 1840, the description of the characters of the various winds expressed in almost the same words as I have used in my papers ever since the

* *Exposition du Système des Vents ; ou, Traité du Mouvement de l'Air à la Surface du Globe et dans les Régions Élevées de l'Atmosphère* (Explanation of the System of the Winds; or, Treatise on the Movement of the Air at the Surface of the Globe and in the Upper Regions of the Atmosphere).

year 1827, and in my Meteorological Investigations in
the year 1837. In this work the rules for both hemispheres are given beside each other. In the passages
relating to the Law of Gyration I shall insert the symbols
relating to the southern hemisphere in parentheses after
those for the northern. I must first insert some general
remarks relating to all observations made at sea.

5. Effect of the Motion of the Ship on Observations of the Law of Gyration.

The evidence adduced from marine observers has referred,
with the exception of the individual observations which
Dumont d'Urville communicated to me, to the general
impressions produced on them by the changes of the
direction of the wind in both hemispheres, and we may
conclude that their observations were made both at sea
and in harbour, as, even in the latter case, the sailor never
neglects to observe the direction of the wind. Observations carried on in this way are really made under very
different conditions, as will be seen if we take an example.
Supposing that at any point on the northern hemisphere
a polar current sets in, and that a ship sails before it
towards the equator, it will observe a gradual shift of the
wind from N. through NE. towards E., while at any
fixed station on the track of that ship the direction of
the vane would remain unchanged, provided that the
point where the current takes its origin remains unaltered.
If, on the contrary, we suppose that the ship sails towards
the point where the current takes its origin, at the same
rate as that point moves away from the fixed station, the
observer on board ship would find that the direction of
the wind remained unchanged, while a shift of the wind
would be observed at the land station. In general an

approach to the equator will accelerate the rotation of
the polar current (from N. through NE. to E. on the
northern, from S. through SE. to E. on the southern,
hemisphere), while an approach to the pole will retard it.
The effect of the former is the same as that which a
removal of the point of origin would produce at a land
station ; the effect of the latter as that of the approach of
that point to the station. Conversely, the approach of the
ship to the equator will retard the rotation of the equatorial current (from S. through SW. to W. on the northern,
from N. through NW. to W. on the southern, hemisphere),
and an approach to the pole will accelerate it.   It seems
most advisable to make these remarks in connection with
the observations of Lartigue, as he points out most distinctly how the same current changes its direction in the
course of its progress.   I have accordingly kept his evidence distinct from that of the other observers.   If we
take the Law of Gyration to mean the regular succession
of the directions of the wind at any one station, as I have
done in all my investigations, it is clear that, accurately
speaking, the sailor, as he is changing his position, does
not observe the same facts as a person placed on land
does, but combines different steps, observed at different
stations, of the same gyration, which passes regularly
through all its steps at each of these stations successively.
We cannot, therefore, obtain from any one log-book results
for the sea which will correspond with those obtained at
the land stations, but must combine the observations recorded in the log-books of several ships which passed the
same point in succession. This is very hard to insure, but
it would be desirable that sailors should make a distinction
between the observations made by them in port and those
made when the ship is under weigh.   It is possible that
the wind might shift against the sun when the ship is

sailing into higher latitudes, without this shift being in reality an abnormal gyration. Such a shift as this is distinguished from those which will be explained hereafter, and which are due to cyclonical movements, by the fact that the change of direction takes place slowly in the former, rapidly in the latter, case.

The results obtained by Lartigue are as follows : —

If in the northern (southern) hemisphere the SSE. (NNE.) wind commences to blow gently, it freshens gradually. If the sky becomes overcast, the wind rises higher, and it shifts round towards SW. (NW.). Rain begins to fall when the wind is between S. (N.) and SSW. (NNW.); the weather then becomes cloudy, and the wind shifts to SW. (NW.) and WSW. (WNW.). It often stays in this point for several days, but usually it shifts quickly to WNW. (WSW.), with occasional heavy showers, which follow each other rapidly. The wind is at its greatest height at this time.

If the WNW. (WSW.) wind continues to blow with the same intensity for a considerable time, it will afterwards shift towards NW. (SW.) or N. (S.), and if the weather then clears up, it is a sure sign that it (the wind) will be able to reach the torrid zone (*sic*).

If the force of the wind decreases after that it has shifted from SW. (NW.) or WSW. (WNW.) to WNW. (WSW.), and the weather does not clear, it will shift back to WSW.* (WNW.), blowing meanwhile from the intermediate points. The WSW. (WNW.) wind, after a time, takes a fresh shift to WNW. (WSW.), and returns again to WSW. (WNW.). If the wind should blow during these oscillations from any point between NW. (SW.) and N. (S.), or between SW. (NW.) and SSE. (NNE.), it may attain the intensity of a storm.

In all cases, immediately before the WNW. (WSW.) wind displaces the WSW. (WNW.), the latter approaches the south point (north point) and increases in intensity, and when the

---

* This is the phenomenon which I have termed the frequent resilition of the wind on the west side.

WNW. (WSW.) begins to blow, and also during the shower, its direction approaches the north point (south point).

Generally, the winds shift from WSW. (WNW.) to WNW. (WSW.), but frequently, in latitudes above the parallel of 40°, they only veer from W. ¼ SW. (W. ¼ NW.) to W. ¼ NW. (W. ¼ SW.). In the neighbourhood of the equator they shift regularly from SW. (NW.) to NW. (SW.), and even to N. (S.).

If the wind begins, after a calm, to move from the pole towards the equator, it assumes in general, in the temperate zone, a northerly (southerly) direction, and shifts gradually to NNE. (SSE.) and NE. (SE.) in its progress southward (northward). It does not attain its greatest intensity until it reaches the torrid zone, and blows from the NNE. (SSE.).

If the wind shifts, in the temperate zone, in this direction, viz. from left to right, like the hands of a watch, it should be remembered that it never attains great intensity excepting at the moment of the shift from WSW. (WNW.) to WNW. (WSW.), and that it never holds long in the NW. (SW.) and N. (S.) points unless in districts where the configuration of the land causes its intensity to increase, as in the Gulf of Lyons and the Gulf of Mexico (the Bay of Rio de la Plata), where it often blows in squalls. If, on the contrary, the wind in the temperate zone shifts in the opposite direction, viz. in the opposite way to the hands of a watch, great disturbances in the condition of the atmosphere may ensue, such as squalls, gales, and hurricanes.

There are, however, some exceptions to this rule. It sometimes happens on the west coast of France, and in very high latitudes, that the wind, after a calm, commences from the S. (N.), then shifts to SE. (NE.), E. (E.), and NNE. (SSE.), without attaining great intensity, and without any change of weather from fine to wet; but if the wind passes from NNE. (SSE.) to N. (S.) and NW. (SW.), it may blow in squalls or even rise to a storm.

The observations which were made by Lartigue himself are given in the 'Annales Maritimes,' 1841, p. 258 sqq.

NORTHERN HEMISPHERE. In the roadstead of Brest I have

observed for many years that the wind shifted from WSW. to
WNW., that the weather was showery when the wind was in the
latter point, and that it became fine when the wind was from
the NW. In Paris I have observed nearly the same facts. After
calms, both at Brest and at Paris, the wind commences to blow
from the S. and SSE. and shifts to SSW. and SW.

Ship *Marengo*, Nov. 6, 1814. Lat. 35° 25′ N., long. 19° 40′ W.
After light winds from the SE., S., and SSW., which lasted ten
hours, the wind blew from the W. and then shifted to NNW.,
where it blew fresh ; farther to the southward we found the
wind northerly, and then, for three days, very fresh from the
NNE.

Nov. 13, 1814. Lat. 29° 25′ N., long. 23° 10′ W.   After
a calm of some hours' duration the wind rose from the SSW.,
then shifted to SW. and WSW.; after two days to W. ¼ NW.
for five hours, then to NW. and N., shortly afterwards to NNE.
and NE.

Frigate *Cybele*, June 17, 1816. Lat. 45° N., long. 48° W.
After a calm for twelve hours a breeze got up from the
SSE., which then shifted to S., SSW., WSW., and finally to
WNW.

June 24, 1816. Lat. 45° 50′ N., long. 55° W.   After a
calm for seven hours a gentle breeze came from the SE., which
freshened gradually as it shifted to SSW. and SW.   On the 25th
the wind was N., then NNE.

Oct. 7 to 9. Lat. 45° 37′ to 50° N., long. 56° W.   Three
times, after a calm, the wind shifted from S. to SW., and then
suddenly to NW.   Between lat. 45° 37′ N., long. 55° 48′ W.,
and Brest, the wind shifted from WNW. to N.

Corvette *La Zélée*, June 29, 1818.   At Rochefort the wind
was N., next morning NE., ENE., E.

Lat. 41° 20′ N., long. 13° 30′ W.   The wind was at NNW.
and NW., and became NNE. and NE. after we passed the
parallel of the Straits of Gibraltar.

Frigate *Clorinde*, Aug. 11, 1821. Lat. 37° 15′ N., long.
14° 25′ W.   A light wind from WNW. veered in twelve hours
through N., NNE., NE., and E.   In lat. 33° N., long. 17° W., it
was NE. and NNE.

Goellette *Lyonnaise,* Feb. 18, 1825.   Lat. 37° N., long. 15° 45′ W.   The wind was WNW. for ten hours, then N.; ten hours later, ENE.   On the 20th it was E., and then SE., in lat. 35° N., long. 18° W.   On the afternoon of the 21st, in lat. 32° 15′ N., long. 20° 50′ W., it was first SSE., then S., SSW., and SW.   On the 22nd it was WNW., NNW., NNE., and then NE. to Teneriffe.

Brig *Alcibiade,* Oct. 17, 1828.   Lat. 41° 10′ N., long. 17° 30′ W. After a calm for two hours a breeze came from SSW.; it then freshened and shifted to SW. and WSW.   After thirty-six hours it shifted from WSW. to WNW. and then to NNW.

Oct. 20.   After a calm for some hours in lat. 37° 15′ N., long. 23° W., the wind was SSW., then it shifted from SW. to NW., and subsequently to NE., E., and SE., until the 25th, in lat. 34° 5′ N., long. 31° W.   It then became SSW., and veered through N. to NNE., NE., E., SE., and SSE.

Oct. 28.   Lat. 33° 27′ N., long. 39° W.   A southerly wind, which was very squally, became first WSW., then WNW., and lastly NE.

Nov. 4.   Lat. 34° 18′ N., long. 64° 40′ W.   The wind was E., then ESE.   In lat. 36° 19′ N., long. 70° 52′ W., it was S., then SW.   On November 6, it was SW., then WNW., and shifted after a violent squall to NNW. and N.

Two similar instances, one between Norfolk and St. Domingo, where the wind shifted from N. to NE., the other between Pensacola and the Havanna, where a northerly wind shifted to NE. and then to E.

Melpomene.   In the instance here quoted the ship sailed from the Straits of Gibraltar to the northward as far as the parallel of 45°, so that the change of wind took place in the retrograde direction, since the ship, in higher latitudes, found a wind which had not undergone so great a deflection, owing to the motion of the earth.   It is probable that if the ship had remained stationary, the wind would not have appeared to have changed, provided that the intensity and the distance of the initial point of the current had remained unaltered.   The wind in lat. 42° N. was ENE.; in 43° 15′ NW., in 44° WNW., and in 45° in the meridian of Cape Finisterre it was between WSW. and W.   It is possible

that this was a case in which a wind blowing in the direction of
a parallel of latitude was deflected in the same way, as, according to Hadley's original principle, it is only possible for winds
to be deflected which make a certain angle with these parallels.

Frigate *Atalante*, Nov. 1833. In lat. 29° 10′ N., long. 27° W.,
a strong NNW. wind, after forty-eight hours, became less violent, and veered to N. and NNE.; then to NE. in lat. 21° N.,
long. 37° W., and at last fell more and veered almost to E.

Jupiter, Feb. 5, 1836 (abnormal). The wind was NNE.,
then N., NNW., and finally NW., while the ship was sailing
westwards. In lat. 42° 55′ N., long. 13° 45′ W., the gyration
was normal, for as the ship went southwards the wind became
first NNE. and then ENE.

Corvette *Caravane*, Sept. 12, 1838 (abnormal similarly).
In the Straits of Gibraltar the wind was NE. and ENE., then
NNW. and NW., in lat. 35° 11′ N., long. 10° 6′ W.; subsequently
W., then SW., and lastly SSW. The normal shift to NNW.
and N. did not take place till October 1, in lat. 20° 41′ N., long.
41° W., viz. SE., S., SW., NW., N.

In addition to the observations which Lartigue himself
made, he quotes the following authorities in confirmation
of his views:—

GREAT NORTHERN OCEAN. *Lapérouse*, pp. 345, 305, 307, 309, 311.

Portlock and Dixon, 1785, p. xx. (two instances); App. xxiv.
(two instances); pp. xxv., xxvii., and xxviii.

Cook, 1779, iv. 514, 515. Indian Ocean, 518.

Ship *La Bonite*, 282, 283.

ATLANTIC OCEAN, SOUTHERN HEMISPHERE. *Bougainville*, 1826,
ii. 155, very remarkable instance; 157, the same.

Duperrey, 13, 1822. ' *Observat. Météorol.*' two instances, 25;
three instances, 117, 119, 121.

PACIFIC OCEAN, SOUTHERN HEMISPHERE. *Lapérouse* from Tal-
caguana to Easter Island. The S. and SSW. winds change to
SSE., SE., and ESE., in proportion as the latitude becomes
lower, 285.

Indian Ocean, Bougainville, 1824, 111, two instances.

New Holland, 139.

From Port Jackson to Valparaiso, 145 (two instances), 147, 149.

From Valparaiso to —— , 15, 153, 155.

Ship *L'Astrolabe,* 1828, 131 (two instances), 137.

Cook, 1773, iv. 231, 490.

Duperrey, ' *Observat. Météorol.*' 25, 31, 49.

Ship *La Bonite,* 1836, 101, 102, 106, 111, 120, 147, 148, 149.

In all the descriptions which have been hitherto given of the fluctuations of the instruments, reference has been made to the direction of the wind as indicated by the vane. I have described (p. 83) the manner in which the currents displace each other, in the same terms as I used in my ' Meteorological Investigations,' * and have explained there how the equatorial current makes its appearance in the upper strata of the atmosphere and displaces the polar current from above, while the polar current, on the contrary, first appears in the lower strata and gradually rises. In this respect the statements of Lartigue agree perfectly with mine, with the sole exception that he uses the term (p. 47) *vents primitifs* (primitive winds) for the current which I have called the polar current, and *vents secondaires* (secondary winds) for the equatorial current. I do not see any reason to change the nomenclature which I have introduced, inasmuch as it is employed in Germany, England, Russia, and also in France, as appears from a complete review of my works by Laugel in the ' *Revue des deux Mondes.*'

An immediate result of this mutual displacement of the winds is that, very frequently, winds which are blowing in opposite directions are flowing one above the other. Should the difference of temperature between the two currents produce a mist at the bounding surface,

* *Meteorologische Untersuchungen.* 8vo. Reimer, Berlin, 1837.

and should this take place within the upper current, we
are able to ascertain the direction of this upper current
by observing the drift of the clouds from below.

There are two distinct causes to which the precipitation
of moisture may be referred (neglecting those due to the
courant ascendant [ascending current]), one of which is
that the equatorial current, in its passage into higher
latitudes, has its temperature lowered ; the other that the
contact of the two currents causes a condensation of
moisture. The first-named of these I have termed the
' precipitations of the current,' the second the ' precipitations of the change of currents.' It is evident that in
the first of these the drift of the clouds corresponds to
the under current, and that, inasmuch as the equatorial
current is a south wind which has received a westerly
deflection proportionate to the distance which it has travelled, the directions of the wind for the west side,
obtained from the drift of the clouds, must exceed in
number those observed on the vane at the surface of the
earth. The opposite must be the case with the east
winds : for, as they bring clear weather, it will be impossible to demonstrate the existence of the current at a
high level when it is prevalent in the lower and upper
strata, inasmuch as the materials by which the current
might be rendered visible below are absent. This shows
us that, although we may admit that the direction of the
wind, which is given by the drift of the clouds, is not
affected by so many of the disturbing actions exerted by
the surface of the earth on the air which is flowing over
it, yet we must not forget that a material complication is
introduced by the fact that the clouds give the direction
at times of the lower, at times of the upper current, while
the vane of the weathercock only indicates the point
from which the under current is blowing.

In this way the results obtained by Bertrand le Doue ('*De la Fréquence comparée des Vents Supérieurs et Inférieurs*' — On the Comparative Frequency of the Upper and Under Currents) are to be explained. He finds from observations carried on by himself at Le Puy for five years, by Quetelet at Brussels for eight years, and by Müller at Görsdorf for four years, the following proportions : —

	Le Puy		Brussels		Görsdorf	
	Upper	Under	Upper	Under	Upper	Under
NE.	81	160	82	84	32	43
E.	9	14	82	121	110	128
SE.	31	122	28	64	66	125
S.	82	117	82	121	85	73
SW.	125	75	244	178	274	210
W.	194	120	282	180	220	217
NW.	232	222	116	87	133	75
N.	246	179	84	65	71	29

With SW. below and SE. winds above, the amount of rain which was collected at Le Puy was 296 millimetres; with SW. above and SE. below, it was 131 : so that in the former case it was greater than in the latter, while the proportion with other winds was the opposite. This is due to the fact that with the winds mentioned (SW. and SE.) the precipitations of the current unite with those of the change to produce the total result.

There is another circumstance affecting this result to which Broun ('General Results of the Observations in Magnetism and Meteorology made at Makerstown in Scotland,' p. 104) has drawn attention. Inasmuch as the earth seeks to impart its own velocity of rotation to the air which is in contact with it, it has a tendency to exert an action in diminishing the deflection which has been

imparted to the current in its passage over parallels of
latitude whose velocity of rotation is continually decreasing.   Broun found the direction of the wind at the
surface of the earth W. 21° S., of the ' under scud current ' W. 7° S., of the Cirrostratus · current W. 2° N.,
and lastly that of the Cirrus W. 9° N.

The same law seems to hold good in the torrid zone, in
the region where the Monsoons prevail, viz., that when
the alternation of the currents takes place, the displacement of one current by its successor from the opposite
point of the compass does not take place simultaneously
in all the strata of the atmosphere.   Le Gentil says
(' Voyage,' i. 485) that, for three or four weeks before
the change of the. Monsoon, the drift of the highest
stratum of clouds was in the opposite direction to that
of the existing Monsoon at the surface of the earth, and
in the direction of the Monsoon which would ensue after
the·change.

THE LAW OF STORMS.

I.—The Storms of the Torrid Zone and their Extension into the Temperate Zones.

THE idea that any considerable diminution in the pressure of the atmosphere must be the result of some unusual disturbance in that medium, offers itself naturally to our mind, and it has, therefore, been long since maintained by those who observed that the weight of the atmosphere varied at different times. Otto von Guericke had attached a scale to the water barometer which he had invented in order to measure these variations, and quotes in chap. xxi. of the ' Mirabilia Magdeburgica,' in Schott's ' Technica Curiosa,' a remarkable observation:— 'In the year 1660 the air became so unusually light, that the finger of the little figure indicated a point below the lowest mark on the glass tube. When I saw this, I said to the persons who were present that, without doubt, a violent storm had arisen somewhere or other. Two hours had hardly passed when that storm burst over the country, although with less fury than it had exhibited on the ocean.' A more modern instance of the same fact is given by the storm of January 17, 1818, whose fearful ravages could be traced by me in the year 1827 in the forests of Prussian Lithuania, nearly ten years after the time when its devastations were felt from the coast of England to Memel, over an area of 1,100 miles in length and 190 in breadth. On January 18, the barometer at Königsberg fell 0·71 in. in eight hours, and between the

3rd and 17th 1·865 in. in all.   In Dantzic it fell 1·598 in.
In Edinburgh also, where the effects of the storm were
such as are usually only produced by electrical explosions, the fall of the barometer was very great.   In
fact, the remark of Otto von Guericke has been so abun-
dantly confirmed by the experience of the two centuries
which have elapsed since his time, that the lettering
attached to our barometers at the present day usually
ends with the words 'stormy.'

The truth of these rules is not confined to the temperate zone.   Scoresby earnestly recommends the use of
the barometer to the seamen whom the whale fishery
attracts to the dangerous waters of high latitudes.   In
consequence of a fall of his marine barometer of 0·825 in.,
on April 5, 1819, in lat. 70° 49' N., long. 70° 15' W., he
was warned of and escaped a storm, which raged for
two days incessantly.   Similarly, several instances are
given from the Trade-wind zone and the district of the
Monsoons, in which an unusual diminution of pressure
preceded West India hurricanes and typhoons.   On
July 26, 1825, the same fact was observed when
Basseterre in Guadaloupe was destroyed by a storm, of
whose violence some idea may be formed from the report
of Gen. Baudrant, in which it is stated that three twenty-
four pounders were blown away by it, and that a piece
of deal board, 37 inches long, 9 inches wide, and $\frac{7}{8}$ inch
thick, was driven through a palm tree 16 inches in
diameter.*   Similar atmospherical conditions accom-

---

* It sometimes happens that whirlwinds of smaller dimensions produce
extraordinary mechanical effects.   Such a storm, whose diameter was only
a quarter or half an English mile, passed, on April 8, 1833, between
Calcutta and the great salt-water lake, about three miles to the eastward
of the city.   It was felt over an area sixteen miles in length, and in the
course of four hours killed 215 persons, wounded 223, and blew down

panied a fall of the barometer of 1·154 in. at St. Thomas, on September 29, 1819. ·On August 2, 1837, at 4 o'clock P. M., the harbour-master of Porto Rico notified to the masters of vessels that they ought to prepare for a storm, as the barometer was falling considerably. At 8 o'clock P. M. it had fallen to 29·601 in. ; at 11 o'clock to 29·300; and ultimately to 28·0 in. The fall was about the same as at St. Thomas, where it fell from 29·930 to 28·064 in. during the same storm. All the precautions were in vain. Not a single one of the 33 ships which were lying at anchor could be saved, for the storm was of such violence that 250 houses were destroyed in St. Bartholomew alone. The ruin at St. Thomas was still more fearful, as the wrecks of 36 ships bestrewed the harbour, and the fort at its entrance was as much shattered as if it had been exposed to a bombardment. In this case, too, some twenty-four pounders were blown away. A large well-built house was torn from its foundations and left standing upright in the middle of the street, while other houses were completely blown down. We read in the 'Annales Maritimes' (ii. 550) the following account of the storm of January 26, 1825, at Guadaloupe:— 'Five ships, which had lain at anchor in the roads of Basseterre, disappeared, and only two of the captains were saved : one of them, Mackeown, saw his brig, after a struggle with the raging sea, carried up by a whirlwind— *faire, pour ainsi dire, naufrage dans les airs* (to be wrecked, so to speak, in the air).' The phenomena which accompany the storms of the Indian Ocean are precisely

---

1,239 fishermen's huts. It forced a long bamboo cane through a wall five feet thick, so as to pierce the facing of the wall on both sides. The editor of the *India Review* remarks that a six-pounder could hardly have produced the same effect.

analogous to those just mentioned.   In the night between
February 28 and March 1, 1818, the barometer fell
at the Mauritius, during the hurricane, to the level of
28·064 in. (reduced to the height at the sea-level).*
It fell almost as low during the hurricane of March
1836.   On the 6th, at 5 A. M., its height was 29·930 in.,
and by 8 o'clock on the morning of the 8th it had
fallen to 28·229 in.   Here, too, the force of the wind
was almost incredible.   The Theatre was a T-shaped
building, and the body of the house, which was 82 feet
in length and 34 in breadth, was torn from the *façade*
on March 1, 1818, and removed to a distance of 5 feet
from its foundations.

Whenever two occurrences constantly take place together, we may, with some probability, conjecture that
they are mutually connected in the relation of cause and
effect.   We cannot decide which of them is the cause
and which the effect, as it is not impossible that they
may both be the results of some third primal cause.   We
are equally unable to decide immediately whether, even
if one of the occurrences may have been directly the
result of the other, the same effect might not have been
brought about in a different manner.

Barometrical minima constantly occur when there is a
considerable disturbance of the atmosphere.   We see, however, that the level of that instrument is frequently very
low at a time when mild spring winds seem to waft us
from the severity of winter into a more genial season.
Observers found it hard to convince themselves that so

* In the account we read, 'Jamais on ne l'avait vu aussi bas.   Plusieurs
personnes crurent que leurs baromètres étaient dérangés, celles qui ne pou-
vaient se méprendre sur la cause de cette dépression, s'attendaient à une
grande catastrophe.'   (It had never been seen so low before.   Many persons
thought that their barometers were out of order, while those who could not
be mistaken as to the cause of the depression expected a great catastrophe.)

gentle breezes could disturb the equilibrium of the atmosphere to the extent that their instruments seemed to indicate, and they consequently attributed the great diminution of pressure to other causes. It is a very natural idea to connect the fearful convulsions of the earth's surface, which take place during an earthquake, with disturbances of the atmosphere, and it has therefore been not uncommonly supposed that the barometer should indicate the occurrence of such a catastrophe in distant countries. This view seemed to be confirmed by the fact that four days after the destruction of Messina, in the year 1783, the barometer exhibited an unusually low level over the whole of Europe. Van Swinden was led from this to believe that a connection existed between the two phenomena. Brandes instituted a comparison of the meteorological observations which were made at the time, as given in the Mannheim Ephemerides, and found that on the 9th of February the amount of depression below the mean level of the barometer was 1·243 in. at Lyndon in Rutlandshire ; 1·199 at Amsterdam and Francker ; 1·132 at Dunkirk ; 1·110 at Middleburg ; 1·088 at Paris ; 1·021 at Laon, Nantes, and Cambray ; 0·932 at Brussels, Chartres, Poictiers, and Rochelle ; 0·888 at Troyes and Montmorenci ; 0·799 at Göttingen, Mayence, Metz, Limoges, and Bordeaux ; 0·710 at Copenhagen, Erfurt, Würzburg, Lyons, Mezières, in Guyenne, and at Oléron ; 0·622 at Spydberga in Norway, Stockholm, Berlin, Vienna, Mannheim, Geneva, and Vienne ; 0·533 at Sagan, Prague, Ratisbon, on Mount St. Gothard, and at Montpellier ; 0·444 at Marseilles and Montlouis ; 0·355 at Buda and Padua ; and 0·267 at St. Petersburg, Mafra, Bologna, and Rome.

It will be seen from this that the barometer was lowest in England and Holland, and this depression decreased gradually as the point of observation approached Italy. This fact seems to show that the phenomena were, in all probability, independent of each other.

If simultaneous observations, like those quoted, are sufficient to prove that the simultaneous occurrence of two phenomena, between which a connection has been supposed to exist, is really only a casual coincidence, inasmuch as the facts are entirely independent of each other, we may naturally expect that, by submitting such observations to a careful examination, we shall be able to arrive at the true cause of the phenomena.

On Christmas Eve, 1821, after a long continuance of stormy weather, the barometer in Europe sank to so low a level, that the attention of all meteorologists was attracted to the circumstance. Brandes accordingly published a request in the scientific journals that all the observations which had been made at the time might be forwarded to him, and embodied the results of his investigation in his 'Physical Dissertation on Sudden Changes which have been observed in Atmospherical Pressure.'[*] He arrived at the conclusion that an unknown cause [†] of diminution of pressure passed over the earth at that time, and that the air flowed in from all sides to the point where the pressure was least. Accordingly, the storm which had resulted had been centripetal ('*vergere*

---

[*] *Dissertatio Physica de repentinis Variationibus in Pressione Atmosphæræ observatis* (Physical Dissertation on the sudden Variations which have been observed in the Pressure of the Atmosphere). 4to. 1826.

[†] *Quæ autem causa fuerit pressionis tam valdè imminutæ, utrum aer prope litora maris Atlantici omnino e medio sublatus fuerit, utrum oceani fauces aperuerint ut aerem haurirent, an imbres, fulminum ̓ vi excitati, massam ejus imminuerint, nemo est, qui dicere possit.* (But there is no one who can say what may have been the cause of so great a diminution of pressure, whether a mass of air in the neighbourhood of the coast of the Atlantic entirely disappeared, or whether the jaws of ocean opened to absorb the air, or showers produced by electricity, lessened its volume.)

procellarum directionem ad idem illud centrum' [that the
directions of the storms tended to one and the same
central point]), and had arisen from an effort of the surrounding air to restore the equilibrium which had been
disturbed at a definite point.

Previous to this, in the year 1820, Brandes had sought
to establish this view, in his history of the weather for the
year 1783, by the comparison of several barometrical
minima similar to the above.   Even these examples are
remarkable for the very slight amount of coincidence
which they exhibit with his views.   One of these storms,
on the night of the 11th of March, travelled in three
hours from Naples to Venice, according to the account of
Toaldo.   The distance is 276 Italian miles, which gives
a velocity of 140 feet per second.   In this case a convergence towards a central point of minimum pressure
in Switzerland is so unlikely, that Brandes himself was
obliged to state, that the current of air which was forcing
its way towards Venice with such extraordinary velocity,
had generated a kind of whirlwind on a prodigious scale,
in which the air had flowed from Marseilles to Corsica, in
order (as he conveniently adds) to join on to the main
current at that point.   Brandes goes on to say, ' Yet these
are only suppositions ; but so much is certain, that, as the
wind was E. at Copenhagen and SE. at Buda, a convergence
took place almost all round the vortex.'   The only confirmation of this is the north wind at Berlin ; so that we
may, with much greater probability, assume the directions,
which he has given, as tangents to circles around the
central point, than as radii towards that point.

The theory which I had formed for myself relative
to the mean atmospherical variations, was that they
owed their origin to the struggle between two currents
which alternately displaced each other at the point of

observation.   It was a necessary consequence of this that the absolute extremes of these fluctuations must be due to the sole predominance of each of these currents.   According-ly a barometrical minimum must be a phenomenon of the southern current: if it be observed simultaneously at several stations, it represents the southern current itself: locally considered, it appears as a stormy passage through the minimum of the wind-rose.   If both of these concur together, there must be a cyclone travelling in the direction of the southern current, *i. e.* from SW. towards NE. In order to establish this view, I submitted to a fresh examination the observations which had been collected by Brandes and others, and proved, in a paper on ' Baro-metrical Minimums,' published in the year 1828, in Poggendorff's Annalen (vol. xiii. p. 596), that all the facts might be simply explained on the assumption that one or more whirlwinds of large dimensions had travelled from SW. towards NE.   I remarked at the same time that in most of the hurricanes of the southern hemisphere which I had examined, I had found the rotation of the whirlwind to be in the opposite direction to that observed in the northern hemisphere.   The instance which I investigated at that time affords a perfect confutation of the theory of convergence to a centre, and I shall therefore repeat here the most important of the quantitative results which I obtained.

The following was the depression of the barometer below its mean level at each station, in inches:—

6 P. M., DECEMBER 25.

Brest. . . .	1·954	Dieppe	1·465	Joyeuse ⎱ Augsburg ⎰ . .	0·799
Helston ⎱ Nantes ⎰ . .	1·687	Strasburg ⎱ Geneva ⎰ . .	0·977	Ratisbon ⎱ Leipzic ⎰ . .	0·710
Gosport	1·509	Bremen			
London ⎫ Haarlem ⎬ . . Paris ⎭	1·332	Zürich ⎫ Göttingen ⎬ . . Bergen ⎭	0·888	Prague ⎫ Breslau ⎬ . Christiania ⎭	0·621

6 P.M., DECEMBER 25 — continued.

Würzburg . . .	0·755	Turin }		Florence . . .	0·293
Peissenberg . .	0·666	Modena } . . 0·444		Rome	0·133
Cracow }		Tilsit }		Molfetta }	
Apenrade } . .	0·577	St. Petersburg } . 0·266		Archangel } . .	0·089
Abo }					

3 A.M., DECEMBER 25.

London . . .	1·954	Heidelberg . .	1·332	Turin	0·799
Dieppe . . .	1·909	Geneva }		Milan }	
Gosport }		Zürich }		Cracow } . . .	0·710
Boston } . .	1·776	Augsburg } . .	1·154	Christiania . .	0·666
Helston . . .	1·687	Berlin }		Abo	0·533
Paris	1·643	Bergen }		Florence }	
Haarlem . .	1·598	Joyeuse . . .	1·110	Rome } . .	0·444
Kinfaun's Castle	1·554	Ratisbon }		Tilsit }	
Strasburg . .	1·465	Gotha } . .	1·066	Molfetta . . .	0·266
Ratisbon }		Leipzic }		St. Petersburg .	0·222
Cologne } .	1·243	Prague }		Archangel . . .	0·111
Göttingen }		Breslau } . . .	0·977		

10 A.M., DECEMBER 25.

Middelburg . .	2·042	Heidelberg }		Cracow }	
Gosport . . .	1·865	Gotha } . .	1·243	Dantzic } . . .	1·021
Haarlem . . .	1·820	Leipzic }		Padua	0·977
London . . .	1·598	Zürich }		Christiania . .	0·843
Helston . . .	1·509	Augsburg }		Florence . . .	0·710
Dieppe }		Vienna } . .	1·154	Tilsit	0·621
Göttingen } .	1·420	Prague }		Rome }	
Bremen }		Breslau }		Molfetta } . . .	0·533
Paris }		Joyeuse }		Abo }	
Strasburg } .	1·332	Innspruck } . .	1·110	St. Petersburg .	0·266
Bergen }		Peissenberg . .	1·066	Archangel . . .	0·111

8 P.M. DECEMBER 25.

London . . .	1·509	Paris }		Padua	0·932
Helston }		Gotha }		Peissenberg }	
Apenrade } .	1·465	Breslau } . .	1·154	Prague } .	0·888
Haarlem }		Christiania }		Tilsit	0·799
Bergen } . .	1·421	Strasburg }		Florence . . .	0·710
Bremen . . .	1·332	Berlin } . .	1·065	Molfetta . . .	0·621
Dieppe }		Cracow }		St. Petersburg .	0·355
Göttingen } .	1·243	Turin }		Archangel . . .	0·111
Dantzic }		Zürich } . .	0·977		
		Augsburg }			

If Brandes, in this case, talks of an absolute disappearance of a definite mass of air, and Meissner, in speaking of the barometrical maximum of the 7th of February, 1821, supposes that air was developed in considerable quantity at some undiscovered point, we may confute both these theories by the consideration that for every barometrical maximum there is a corresponding minimum in an adjacent district, and vice versâ. Thus, on the 22nd of January, 1850, the barometer at Königsberg reached its maximum height (0·888 in. above the mean for the month), and on the same day a barometrical minimum (0·736 in. below its mean level at North Salem) was observed in the states of New York and Rhode Island. The minimum (1·794 in. below the mean level) at Christiania, on the 6th of February, 1850, was accompanied by a maximum (0·834 in. above the mean) at Cambridge, near Boston (U. S.), and the minimum (1·660 in. below the mean) at Upsala, on the 1st of January, 1855, by a corresponding maximum (0·382 in. above the mean) at Lisbon. These instances will be treated of subsequently.

The theory of absorption receives no confirmation from the observed directions of the wind.

If a convergence towards such a point were to take place, we should find that equilibrium would exist between the particles of air along a line of equal diminution of pressure, and the paths of the wind would be, on the whole, normals to this line. On the hypothesis that the phenomena are the results of a rotatory motion, the paths of the wind will correspond with the lines of equal diminution of pressure, and consequently be at right angles to the directions given by the absorption theory.

The observations which have been cited show that the minimum travelled from the coast of France towards the south-west point of Norway, pretty nearly from Brest to Cape Lindenaes. We must ascertain accordingly how the directions of the wind are related to this minimum, which is moving as has been described ; whether they converge towards it, or are tangential to circles which have the point of minimum pressure as their common centre, and advance with it gradually.   The simplest test of this is to lay down on four maps the position of the minimum at the four periods quoted (viz.: 6 P. M. on the 24th, and 3 A. M., 10 A. M., and 8 P. M., on the 25th of December), and then mark on these maps the directions of the wind which were observed simultaneously at the several stations. If we find that the arrows thus marked on the maps represent tangents to concentrical circles, we may assume, conversely, that the circles actually existed, and then compare the directions obtained on that hypothesis with those given by the observations.

The path of the minimum is from Brest to Cape Lindenaes; so that we shall find France, Italy, Germany, Denmark, and Russia on the south-east side of the path of the storm; Ireland, Scotland, and Iceland on the north-west side; and England nearly in the centre.

If there were a centripetal convergence, the passage of the storm, at a station situated to the south-east of the path, will produce a rotation of the vane in a regular succession from ENE., through E., ESE., SE., SSE., S., towards SSW.; and its passage, at a station situated to the north-west of the path, from NNE., through N., NNW., NW., WNW., W., towards WSW.

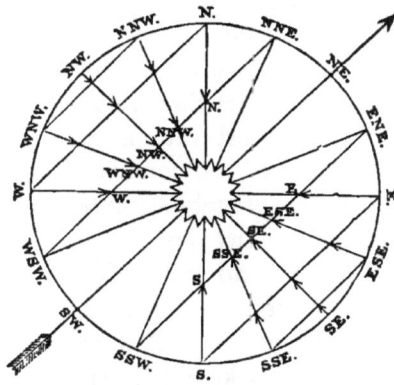

If the storm be a cyclone, revolving in a direction opposite to that of the hands of a watch, the wind will shift on the south-east side, from SSE., through S., SSW., SW., WSW., W., towards WNW., and on the north-west side, from ESE., through E., ENE., NE., NNE., N., towards NNW.

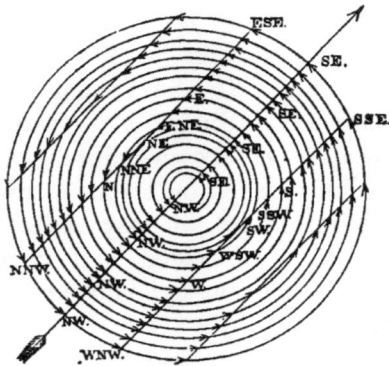

At the stations situated on the path of the storm, the wind will shift suddenly from NE. to SW., according to the first theory; from SE. to NW., according to the second.

According to each of these views, we should expect to find a shift of the wind *with* the sun on the south-east, and *against* the sun on the north-west, side of the path of the storm; and in each case there would be a lull in the centre succeeded by a sudden shift of the wind into the opposite quarter. The difference between the two would be, that the commencement and end of the rotation would be fully 90 degrees distant, in the one case, from what they would be in the other.

All the observations quoted in the paper which has been referred to, confirm the second theory completely, and as completely confute the first. The shift of wind did not in any instance commence with NNE., N., or NNW., and end with WSW., W., or WNW.; but, instead, it everywhere commenced with ESE. and SE., and ended with SW. and W. This was universally true for Germany, Italy, Denmark, and Russia. In England the wind was E., not N., before the minimum. In France it was in general SW. In Iceland it was first NE., then N., precisely the

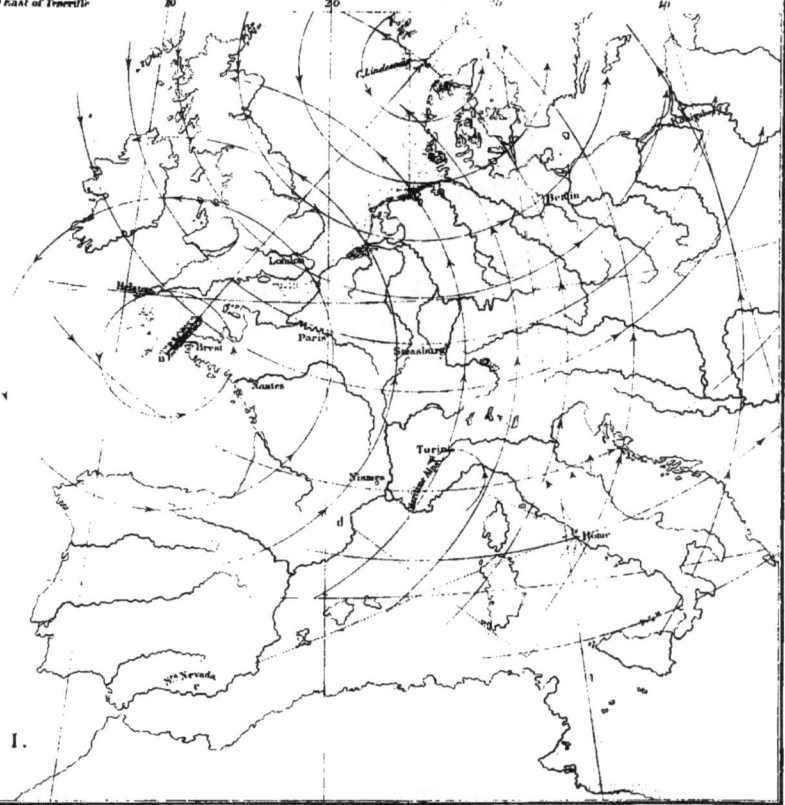

Chart I.

direction which would be expected, at this distant station, on the hypothesis of a rotatory motion.

In my account of the storm which first appeared in the year 1828, in the thirteenth volume of Poggendorff's Annals, in which my object was to give the purely empirical statement of the facts, unmodified by any theoretical considerations, and in my 'Meteorological Investigations' (1837), I have represented the directions of the wind by feathered arrows, the shift of the wind by plain ones. Before the occurrence of the minimum the shift of the wind, at all stations situated along a line perpendicular to the path of the storm, was from SE. towards SW.; at the time of the minimum the wind was SW.; after the minimum the shift was from SW. to NW. In the north-west corner of the diagram an arrow directed from NE. to SW. gives what I call the 'probable direction of the current in North America.' If these be changed for tangents to the circle, we obtain on the right side SW. before the minimum, NW. after it; and on the left-hand side, SE. before, NE. after the minimum. This indicates a cyclone revolving in the direction S.E.N.W., opposite to that of the hands of a watch. According to the practice which has now become general, I have substituted for this purely empirical representation the actual circle.

I have represented on Chart I. this cyclone, from the beginning to the end of its progress, as traced by the observations, in order to facilitate the comparison of the observations with the theoretical assumption. At the places where the cyclone impinged on the mountains of Spain and on the Alps, I have assumed these points as centres of new cyclones. The results of the observations give, as the direction of the wind —

1. On the north-west side of the storm.
    Naes in Iceland, NE., N.

2. Near its centre.

Helston, E., minimum W.
London, SE., min. NW.
Owen's Row, near Islington, SE., min. NW.
Cambridge, SE., min. W.
New Malton, storm from south.

3. On the south-east side of the storm.

Boulogne-sur-Mer, SSE., S., ENE., min. WNW.
Paris, S., min. WSW.
Joyeuse, a storm from south, min. S.
Nismes, S., min. SW., NW.
Vivarais, SE., min. SE.
Strasburg, SE., E., S. min.
Haarlem, SE., ESE., SSE., min. SSW., SW.
Schwelm, S., min. SW.
Cologne, SSE., SE., min. S., WSW., SW.
Coblentz, SW., min. S., SW.
Salzuflen, SE., min. S.
Wetzlar, SSE., min. SSW., SW.
Minden, SE., min. S.
Carlsruhe, S., min. SW.
Göttingen, SE., SSE., min. SW.
Ratisbon, E., SE. min.
Augsburg, SW., min. W.
Quedlinburg, E., min. SW.
Zellerfeld, S., min. W.
Leipzic, SW., min. S.
Zschoppau, min., with SW.
Annaberg, SE., min. SW., W. } disturbing influence of
Prague, W., min. SW., W.   } mountains lying to the
Breslau, SW., min. S.   } south.
Leobschütz, S.
Dantzic, S., min. S.
Königsberg, SE., min. W.
Tilsit, SW., SE., min. W.
St. Petersburg, SE., E., SE., SSE., min.
Archangel, SE., min. S.

Geneva, SE., min.
Zürich, E., min. SE., W.
St. Gall, SE., SSE., min. SE.

4. Modified cyclones on the southern slopes of the Alps.

Turin, W., SW., min. E., NE.
Milan, W., SW., min. W., SE.
Pavia, SE., min. SW.
Modena, SE., min. SW., W.
Padua, W., S., min. N.
Florence, S., SSW., min. SW.
Rome, SSE., S., min. SSE., S., SSW.
Molfetta, SE., S., min. SSW.

This was the first instance that I investigated, and in it all the continental stations lay on the south-east side of the path of the storm in its passage from SW. to NE. At that time I did not know of the observations at Naes, in Iceland, so that I was obliged to state that the NE. wind in North America, which resulted from my theoretical considerations, was only a 'probable' direction. Soon afterwards, I myself observed, at Königsberg, a retrograde rotation. This is alluded to in the paper to which I have referred, in which (p. 613) I say, 'After a minimum of 28·987 in., on the 1st of November, 1827, the wind on the 2nd and 3rd shifted, while the barometer rose, through E. towards NE. On the same day there was a northerly storm in the German Ocean, which might have been foretold from the fact of the exceptional rotation here (i. e. at Königsberg).'

Both the theories which have been contrasted with each other above, have been defended with great vigour of late years. On the one side, Mr. Redfield, after a very careful investigation of the phenomena which accompany the storms which are so frequent on the coasts of the

United States, has arrived at precisely the same conclusions
as I did for Europe; on the other, a champion has come
forward to defend the view propounded by Brandes, in
the person of Mr. Espy of Philadelphia. The tornado of
the 19th of June, 1835, in New Brunswick, led Mr. Espy
to assume the existence of centripetal storms. After this
storm Messrs. Bache* and Espy examined the directions
in which the trees lay which had been blown down in a
wood over which it had passed, and found that all lay
with their tops pointing to one central point. Those on
the west had fallen eastwards; on the north, southwards;
on the east, westwards; and on the south, northwards.
In contradistinction to these statements, Lewis Back, an
eye-witness of the storm, maintains that it was an
unmistakable whirlwind; and asserts, that it would be
impossible to deny this, if the observer had not been
prejudiced in favour of some preconceived theoretical
explanation. Piddington † denies the possibility of forming
a correct judgement on the direction of the wind on land,
unless in great plains, or on low islands. In his Hornbook
(p. 25) he says, 'It seems to me, speaking as a sailor, sheer
nonsense to discuss the question of how *the* wind blows,
inland, in reference to any theory which must depend
upon the direction of winds for short periods and in
storms.' According to Mr. Espy, the cause of the convergence of the currents to a centre is the latent heat
which is set free by the condensation of aqueous vapour
into a cloud, by means of which the air in which that
vapour had been suspended is dilated six times as much

---

* 'Notes and Diagrams illustrative of the Directions of the Forces acting
at and near the Surface of the Earth in different parts of the Brunswick
Tornado of June 19, 1835.'

† Piddington, 'Sailors' Hornbook for the Law of Storms.' 8vo. 3rd edit.
London: Williams & Norgate, 1860.

as it loses in volume by the condensation of the vapour. According to this idea, the air ascends with a velocity of 364 feet per second, and at the level at which the hail clouds lie, it still exerts a pressure of 120 lbs. per square foot. Such a force as this would be sufficient to lift a cubical block of ice, measuring 18 inches each way, or even to carry up an elephant.* Mr. Espy has explained his views more fully in works published in 1841† and 1857‡, to which I must refer the reader. Those of Mr. Redfield's papers which have reached me are the following :—

'Remarks on the prevailing Storms of the Atlantic Coast. Silliman, Americ. Journ. 20, No. 1.'--'Hurricane of August 1831. (To the Editor of the Journal of Commerce.)' 'Observations on the Hurricanes and Storms of the West Indies and the Coast of the United States (Blunt's American Coast Pilot, 12th edit.)'—'On the Gales and Hurricanes of the Western Atlantic. Sill. Journ. 31, No. 1.'—'Meteorological Sketches by an Observer. Sill. Journ. 33, No. 1.'—'Remarks on Mr. Espy's Theory of Centripetal Storms, including a Refutation of his Positions relative to the Storm of September 3, 1821 ; with some Notices of the Fallacies which appear in his Examinations of other Storms (Journ. of the Franklin Institute).'—'On the Courses of Hurricanes, with Notices of the Typhoons of the China Sea, and other Storms. Sill. Journ. 35, Nov.'—'The Law of Storms. New York Observer, January 18, 1840.'— 'Whirlwinds excited by Fires, with farther Notices of the Typhoons of the China Sea. Sill. Journ. 36, No. 1.'—'On Whirlwind Storms, with Replies to the Objections and Strictures of Dr. Hare. New York, 1842, 8vo., 65 pp.'—'Notice of Dr. Hare's

* 'Theory of Rain, Hail, and Snow, Water-spouts, Land-spouts, Variable Winds, and Barometric Fluctuations.' Philadelphia, 1836, 8vo. And, 'Examination of Hutton's, Redfield's, and Olmsted's Theories.'

† 'The Philosophy of Storms.' London, 1841, 8vo.

‡ 'Message from the President of the United States communicating the Fourth Meteorological Report.' Washington, 1857, 4to.

"Strictures on Prof. Dove's Essay on the Law of Storms." (In reply to Robert Hare's "Objections to Mr. Redfield's Theory of Storms, and Strictures on Prof. Dove's Essay on the Law of Storms.")'—'On the First Hurricane of September 1853 in the Atlantic, with a Chart, and Notices of other Storms.'—'On Three several Hurricanes of the Atlantic, and their Relations to the Northern of Mexico and Central America; with Notices of other Storms. New Haven, 1846, 118 pp.'—'Observations relative to the Cyclones of the Western Pacific, 1857.'

A complete list of Redfield's works may be found in Olmsted's 'Address on the Scientific Life and Labours of William C. Redfield.' New Haven, 1857, 8vo.

The materials which were collected with such care by Mr. Redfield have been considerably increased by the publication of a magnificent work* on the subject by the late Major-General Reid, who was at the time Governor of Bermuda. This author has arrived at exactly the same results as Mr. Redfield. They have come to these conclusions, entirely independently of my previous publications, as I have ascertained by personal communication with them. The works of Redfield and Reid, in addition to the further confirmation of the theory of rotatory motion, contain some very important observations, and the credit of establishing these is due to them alone. The most important of these discoveries, that of the motion of the cyclone within the torrid zone from SE. towards NW., before its change of path at the outside limit of the Trade-wind, is due to Mr. Redfield

* 'An Attempt to develope the Law of Storms by means of Facts arranged according to Place and Time, and hence to point out a Cause for the Variable Winds, with the view to practical use in Navigation.' Illustrated by Charts and Woodcuts. London, 1838, 8vo.—'William Reid. The Progress of the Developement of the Law of Storms and of the Variable Winds, with the practical application of the subject to Navigation.' London, 1849, 8vo. 424 pp.

alone.   In fact, the position of an observer in America is particularly favourable for the solution of this part of the problem, inasmuch as the hurricanes which sweep the coasts of the United States traverse, in the tropical portion of their course, the West Indian Islands, where their extraordinary character has given rise to the title, ' West India Hurricanes.'   In the case of the cyclones which are felt in Central Europe, it is rarely possible to trace the tropical portion of the course.   This gives us a satisfactory proof that we shall avoid one-sided views of natural phenomena, the greater be the area over which our observations are extended.   As, in the science of astronomy, the geographical latitude of an observatory exercises an important influence on the problems to be solved there, so, in the science of meteorology, it is equally important to know the position of the station at which the solution of a problem has been undertaken.   I shall now seek to compare the observations of Redfield and Reid with the theory of a rotatory motion.   When I published my first papers on the winds, I was disposed to refer the Law of Gyration as well as the rotatory motion of storms to the mutual interference of two currents of air which alternately displaced each other in a lateral direction.   A closer examination of the phenomena showed me that the Law of Gyration depended on more general principles, and that it was a simple and necessary consequence of the motion of the earth on its axis.*   This generalization of the principle of Hadley's Trade-wind theory explained perfectly all the rules which had been ascertained for the non-periodical fluctuations of the instruments, and enabled us to foretell what they would be in the southern hemi-

---

* *Ueber den Einfluss der Drehung der Erde auf die Strömungen ihrer Atmosphäre.*   (On the Influence of the Rotation of the Earth on the Currents of its Atmosphere.)   Pogg. Ann. xxxvi. p. 321, 1835.

sphere. It did not, however, explain the rotatory motion
of storms, to which idea I had originally allowed too
great an extension, in the introduction to my first investigations (Pogg. Ann. xiii. p. 597), where I said, ' That, in
general, all storms are rotatory is an ascertained fact
which will be confirmed by every seaman.' I fell into
an error, which few of those who have investigated
storms have escaped, which consisted in referring all the
rotations of the vane which may be observed, to cyclonical
movements. I was, accordingly, compelled to retain my
former theoretical explanation, in my ' Meteorological
Investigations ' (1837), in which work I combined into a
whole all the papers which I had published up to that
time. The reason of this was, that the empirical facts
had received entire confirmation, while their connexion
with the principle of the general theory was by no means
so evident. It was not until later* that I was able to
supply this deficiency, by proving that a cyclonical
movement was produced whenever the interposition of
any obstacle interfered with the regular change in the
direction of the wind, which is due to the rotation of the
earth, and consequently interfered with the regular rotation of the vane at any station. From the investigations
of Redfield and Reid we ascertain the following facts.

1. The storms which arise in the torrid zone, preserve
the original direction of their progression (viz., from SE.
to NW.) unchanged, as long as they continue within that
zone ; but as soon as they enter the temperate zone they
turn and move from SW. to NE., almost at right angles
to their former path. The corresponding storms in the
southern hemisphere, which move from NE. to SW. between the tropics, undergo a change of direction, analo-

* ' Reports of the Royal Academy of Sciences of Berlin,' 1840, p. 232.
Pogg. Ann. lii. p. 1.

gous to that above described, on their entrance into the temperate zone, and move from NW. to SE.

2. The diameter of the cyclone, which increases very slowly within the torrid zone, increases to a remarkable extent as soon as the change of direction takes place.

·CHART II.

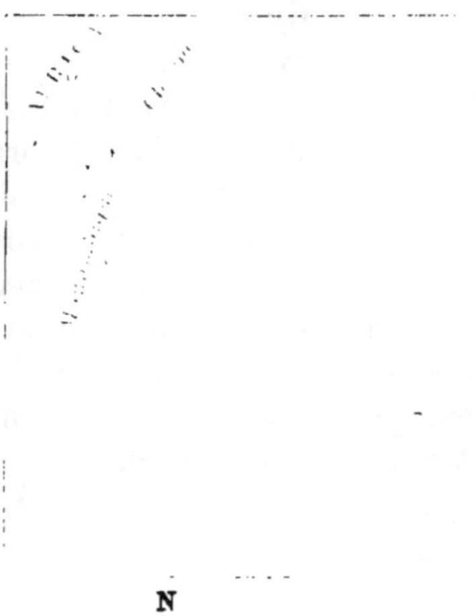

Two charts, which have been borrowed from General Reid's work, and reduced in scale, will serve as instances of the entire phenomenon as presented in each hemisphere. They represent the West India hurricane of the middle of August 1837 (Chart II.), and the Mauritius storm of March 1809 (Chart III.).

CHART III.

N

In order to exhibit the course of the storms by means of numerous examples, we have added to these a chart constructed by Redfield, which gives the course of several of these storms (Chart IV.).

Two of these storms were confined to the torrid zone, and consequently preserved a straight course. One of these was on June 23, 1831, and advanced from Trinidad, by Tobago, Grenada, and the middle of Yucatan, to the neighbourhood of Vera Cruz. The other was on August 12, 1835, and passed from Antigua, by Nevis, St. Thomas, St. Croix, Porto Rico, Hayti, and Matanzas in Cuba, to Texas.

Eight of the storms travelled outside the limits of the torrid zone, and their course was as follows:—

That which devastated Barbadoes on the night of August 10, 1831, reached Porto Rico on the 12th, Aux Cayes and Santiago de Cuba on the 13th, Matanzas on the 14th, the Tortugas on the 15th, the Gulf of Mexico on the 16th, and lastly, Mobile, Pensacola, and New Orleans on the 17th; so that it traversed a space of 2,000 nautical miles in about 150 hours, and consequently travelled with a velocity of $13\frac{1}{3}$ miles per hour. Its direction between the tropics was N. 64° W.

The storm which commenced in the neighbourhood of Martinique on August 17, 1827, reached St. Martin and St. Thomas on the 18th, passed to the north-east of Hayti on the 19th, reached Turk Islands on the 20th, the Bahamas on the 21st and 22nd, the coast of Florida and South Carolina on the 23rd and 24th, Cape Hatteras on the 25th, Delaware on the 26th, Nantucket on the 27th, Sable Island and Porpoise bank on the 28th: and thus travelled 3,000 nautical miles in eleven days. Its direction within the tropics was N. 61° W.; and, on the other hand, in lat. 40°, N. 58° E.

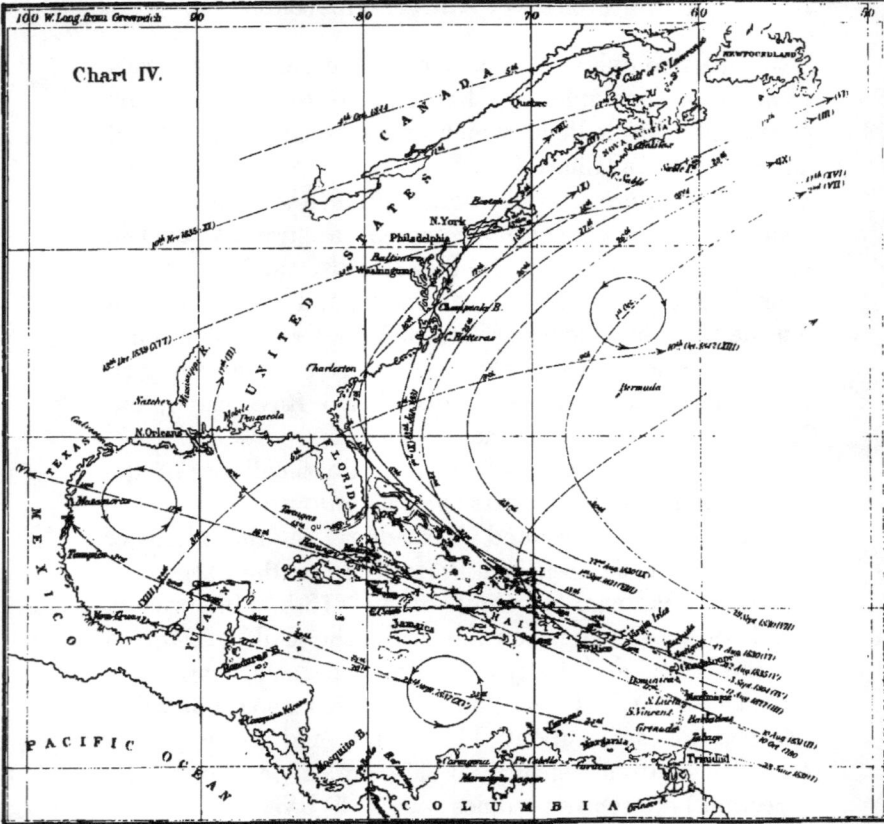

Chart IV.

The storm which commenced in the neighbourhood of Guadaloupe on December 3, 1804, reached the Virgin Islands and Porto Rico on the 4th, Turk Islands on the 5th, the Bahamas and the Gulf of Florida on the 6th, the coast of Georgia, South and North Carolina, on the 7th, Chesapeake Bay, the mouth of the Delaware, and the circumjacent states of Virginia, Maryland, and New Jersey, on the 8th, Massachusetts, New Hampshire, and Maine on the 9th. It moved with considerable velocity, as in its curvilinear course from the Leeward Islands it travelled 2,200 nautical miles in six days, and consequently advanced at the rate of $15\frac{1}{2}$ nautical miles per hour.

The storm which swept close by the Leeward Islands in August 1830, reached St. Thomas on the 12th, was in the neighbourhood of Turk Islands on the 13th, at the Bahamas on the 14th, at the Gulf and on the coast of Florida on the 15th, along the coast of Georgia and the Carolinas on the 16th, on the 17th on those of Virginia, Maryland, New Jersey, and New York, on the 18th at George's bank and Cape Sable, and on the 19th at the Porpoise bank and on that of Newfoundland ; its progress was, therefore, at the rate of 18 nautical miles per hour. If we now assume the real velocity of the wind, in its whirling motion, to be five times greater than that with which the cyclone travels, we find a motion of the air through 18,000 nautical miles in seven days.

The most easterly storm was that of September 29, 1830. It commenced to the north of Barbadoes, in the 20th degree of latitude, recurved towards the north in 68° W. longitude and 30° N. latitude, and passed to the westward of the Bermudas on towards the east end of the bank of Newfoundland, which it reached on October 2.

A very violent storm, of much smaller dimensions, was

that which commenced at Turk Islands on September 1, 1821, was to the north of the Bahamas on the 2nd, on the coast of the Carolinas early on the 3rd, and then at a later hour on the same day on the coast of New York and Long Island, while on the following night it swept over the states of Connecticut, Massachusetts, New Hampshire, and Maine, thus traversing 1,800 nautical miles in 60 hours. Its mean velocity consequently amounted to 30 nautical miles per hour.

The storm of September 28, 1838 had an altogether similar character to the last. On the contrary, that of the 22nd August progressed very slowly. It commenced to the north of Porto Rico, in 22° N. latitude, and advancing in a direction parallel to the coast of North America, did not reach the bank of Newfoundland until the 27th.

At times the storm does not reach its full intensity until it enters the temperate zone, and in this case in particular, we can obtain accurate information about this part of its course. The storm of November 10, 1835, was of this character, it was the most northerly of all, and passed over Lake Erie and Lake Ontario to Prince Edward's Island in the Gulf of St. Lawrence.*

* Inasmuch as the direction in which the cyclone as a whole moves is quite distinct from that in which the wind, in its rotatory motion at any place, blows, it is easy to see that mere local observations may lead a person to most incorrect conclusions. Thus, Raynal, in his *Histoire Philosophique et Politique des Deux Indes* (Philosophical and Political History of the Two Indies), vol. v. p. 72, says that the most accurate observers had arrived at the conclusion that the storms which from time to time devastated the West India Islands could only have come from the NW., and reasons accordingly that they must have come from the mountains of Santa Martha. The fact is, that the direction described merely indicates that these islands lie on the south side of cyclones which pass by from E. to W., rotating in a direction opposite to that of the hands of a watch, and is in accordance with the observations quoted above. On the title-page of the sixth volume of this work there is a spirited sketch of a West India hurricane.

It has further been observed that in general the velocity of progression of the centre increases as soon as the storm changes its course at the edge of the torrid zone. The reason for this will be assigned when we come to the theoretical consideration of the origin of the storms. Thus the storm of August 30, 1853, travelled 7,276 English miles in about 12 days, a distance which gives a mean velocity of progression of 26 miles an hour; and after it passed by the bank of Newfoundland the velocity increased to 50 miles an hour.

As regards the periods at which these storms are most prevalent, Poey ('A Chronological Table of Cyclonic Hurricanes which have occurred in the West Indies and in the North Atlantic from 1493 to 1855 '*) finds the following proportions out of 365 storms : —

No. of Storms		No. of Storms	
In January . . . 5		In July 42	
February . . . 7		August . . . 96	
March . . . 11		September . . . 80	
April 6		October . . . 69	
May 5		November . . . 17	
June 10		December . . . 7	

From this it is evident that they are most frequent in August and September.

I find in the 'Bombay Times' of November 28, 1854, the following table of the distribution throughout the various months of 85 storms, which were observed between the equator and lat. 34° N. : —

January . . . 1	May 14	September . . 11
February . . 2	June 6	October . . . 17
March . . . 4	July 3	November . . 11
April 9	August . . . 5	December . . 5

This shows us that the maxima occur during the months

---

* This work contains a very perfect list of the literature of cyclones, under the title, 'A Bibliographical List of 450 Authors, Books, and Periodicals, where some Interesting Accounts may be found, especially on the West and East India Hurricanes.'

in which the Monsoons change, and that the absolute
maximum is observed at the change from the SW. to the
NE. Monsoon.

As regards the velocity of progression, it appears that
this is occasionally so slight in the case of the Typhoons,
that Capper * seems to have considered them to be
stationary and not progressive.

The most remarkable instance of slow rate of motion
of a cyclone is given by Piddington in the 13th 'Memoir
with reference to the Law of Storms, being the "Charles
Heddles" storm of the Mauritius from the 22nd to the
27th of February 1845.' The progressive motion during
each of these five days respectively was 70, 100, 115, 89,
85 English miles. This gives an average motion of 92
miles per day, or 3·8 per hour ; while the ship itself, at a distance of about 50 miles from the centre, was carried round
and round the cyclone in such a manner that she sailed
1,300 miles, and yet found herself after the five days at
a distance of only 354 miles from the port from which
she had started, as the storm had made five complete
revolutions with her. This is, as Piddington says, a sort
of sea-romance, or, to speak like a good Mussulman, it
is something which might have happened, if it had so
pleased the Prophet.

In the year 1840 I published † an attempt to explain
the origin of these phenomena, which is as follows :—

Let $a\,b$ represent a series of material points, in the
northern hemisphere, parallel to the equator, which are
set in motion, by any cause, towards the pole in the
direction $a\,c$. If the space $d\,b\,h$ were a vacuum, these
points would move to the position $g\,h$, inasmuch as they

---

* Observations on the Winds and Monsoons. London, 1801.
† *Berichte der Berliner Akademie.* (Reports of the Academy of Berlin.)
1840, p. 232. Pogg. Ann. lii. p. 1.

travel from greater to smaller circles of latitude. If, however, the space $d\ b\ h$ contain air which has not been set in motion in the same direction, the particles at b in their motion towards d will constantly be brought into contact, in that space, with par-
ticles of air which are rotating with a less

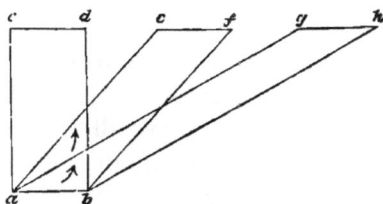

velocity than they are themselves, and, consequently, their velocity of progression towards the east will be diminished. For this reason the point b will move to f, instead of to h. The particles at a are differently circum-stanced, as they are in contact, in the direction of b, with particles which possessed originally a velocity of rotation equal to their own, and, accordingly, their motion takes place as if *in vacuo* : i.e. they move towards g. It follows from this that if $a\ b$ be a mass of air travelling from south to north, the direction of the motion of the air on the eastern side of the mass will be much more nearly that of a south wind, than that of the air on the western side, where it is more of a west wind, and the effect of this difference will be to generate a tendency to a cyclonical movement in the direction S., E., N., W. Such a tendency as this could not exist if the space $d\ b\ h$ did not contain a resist-ing medium, and will therefore increase in proportion as this resistance interferes with the force which ought to deflect the wind towards the west point. Hence the storm will rotate with greater velocity the less the path along which it travels varies from the original direction in which the motion took place (i.e. towards c.) In the zone of the Trade-winds the space $d\ b\ h$ is filled with air, which is moving from NE. towards SW. In this case the greatest resistance which is possible will be presented to the motion of the mass *a b*, and it will become possible that the particles at *b* should be prevented to such an extent from taking the direction of a west wind, that they will actually move towards *d*, while particles at *a* tend to move towards *g*. Accordingly, in this district, the violence of the rotatory motion will be greatest, but the cyclone will advance in a straight line, and its diameter will not increase. As soon as such a storm enters the temperate zone the circumstances are altered, the space *d b h* is filled with air which is already in motion from SW. towards NE. The resistance which had been experienced by the particles at *b*, either undergoes a sudden and considerable diminution, or else disappears entirely: the direction of motion along the line *b d* is changed for one along *b h*, so that the path of the cyclone changes so as to be nearly at right angles to its former course ; and its diameter increases rapidly, owing to the disappearance of the obstacles which produced a difference between the motions of the particles at *a* and *b* respectively.

The explanation in the case of the southern hemisphere follows from the same principles, as may be seen from the diagram. The direction of the rotation of the cyclone

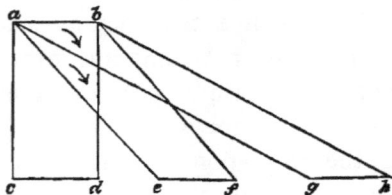

is the reverse of what it is in the northern hemisphere, and the change of the direction of the path at the edge of the torrid zone is analogous.

The preceding explanation of the origin of the rotatory motion is of course only applicable to the cases in which large masses of air of considerable lateral extension are set in motion : smaller whirlwinds (trombs), waterspouts, &c.,

arise from other causes, and we may therefore expect that, as regards the two hemispheres, their motions will exhibit neither conformability to a law, nor a distinct contrast. In fact, Gen. Reid saw, from the Government House at Bermuda, a waterspout which was rotating in a direction contrary to that of the great cyclones, while Mr. Redfield observed a small tornado which rotated in the same direction as the cyclones. The observations of Akin, at Green Bush near Albany, of Dwight, at Stockbridge, Mass., and of Dr. Cowles, at Amherst, on the violent whirlwinds which arise during forest conflagrations when the air has been calm, prove that a strongly developed *courant ascendant* is capable of generating a rotatory motion in the surrounding air.

The West India hurricanes arise at the inner edge of the zone of Trade-winds, in the so-called Region of Calms, where the air ascends and flows away in the upper strata in the direction opposite to that of the Trade-wind. This renders it probable that the primary cause of the cyclones is the intrusion of a portion of this upper current into that which lies underneath it. The reason that the cyclone, in the earlier portion of its course, travels from SE. towards NW., may probably be assigned by the following consideration. According to our theoretical explanation of their origin, a direction similar to that described will be most favourable to the generation of a cyclonical movement. If, as may sometimes happen, the first impulse takes place in the direction from SW. to NE., the north-east Trade, which is blowing in the contrary direction, will offer the same amount of resistance to all portions of the line (the front of the mass of air) which is advancing, and no tendency to a cyclonical movement can arise.

It is well known that, during the eruption of Coseguina on January 20, 1835, when the isthmus of Central

America was disturbed by earthquakes, volcanic ashes were carried by the upper counter-Trade-wind not only as far as Kingston in Jamaica, a distance of 800 English miles against the direct Trade-wind, but also 700 English miles to the westward, where they fell on the ship Conway in the Pacific Ocean. This fact proves that, in the higher regions of the tropical atmosphere, the air does not always move regularly from SW. towards NE., but that the regularity of this movement is interrupted by currents which flow from E. to W. It appears that there is some connection between these facts and the fall of dust, which has come from Africa, in the North Atlantic Ocean, an occurrence which is frequently observed. We must attempt to ascertain the cause of such abnormal currents.

Their origin is to be found in the tables of barometrical fluctuations which have been given (see pp. 54 sqq.). If we compare the annual barometrical curves for Hobarton, Port Jackson, Cape Town, Rio Janeiro, Santiago, Monte Video, and Buenos Ayres with those for Northern Asia, Eastern Europe, and Hindostan, which are represented on pp. 61, 62, we see that the pressure in the southern hemisphere increases at the period at which that in the northern decreases. The most obvious explanation of this would be to suppose the existence of a periodical transfer of the air from the one hemisphere to the other. However, we find that the mean annual tension of aqueous vapour at Barnaoul is 0·191 ins., while at Banjoewangie it is 0·897 ins., so that the supposition of an exchange would be a tacit assumption that air can be converted into water, and *vice versâ*. We are therefore compelled to return to the action of the dry air. A glance at the tables (pp. 52—60) will show that the aggregation of air in the southern hemisphere is insufficient to explain the rarefaction on the northern. It is therefore clear that the air must flow away in a lateral direction on this hemisphere, for aqueous vapour might be condensed during the motion of the air away from Asia to the south, whereas air can never be got rid of in a similar manner. Sitka, in Russian America, affords a proof that the air does really flow from Asia to North America, for at this station the pressure of dry air increases from winter to summer. This is probably the case in Europe also, as the rarefaction in summer is not observed here. We find, however, that the rarefaction extends to the North of Africa, while the barometer remains at an unusually high level at the Canary Islands. The high reading of the barometer is also very remarkable throughout the North Atlantic Ocean, as may be seen from the subjoined table, representing the barometrical conditions of the north-east and of the south-east Trade-winds respectively.

TRADE-WIND ZONE.

Latitude	North-East Trade-Wind Barometer	South-East Trade-Wind Barometer	Difference
50°—45°	30·060		
45 —40	30·078		
40 —35	30·125		
35 —30	30·211	30·052	+ 0·159
30 —25	30·150	30·095	+ 0·055
25 —20	30·081	30·102	− 0·021
20 —15	30·018	30·060	− 0·042
15 —10	29·965	30·028	− 0·063
10 — 5	29·923	29·981	− 0·058
5 — 0	29·915	29·941	− 0·026
0°—35°	30·038	30·037	

Let us now conceive that the air which ascends over Asia and Africa flows away laterally in the upper strata of the atmosphere; a fact which is proved evidently by the dust which has been observed to fall in the Atlantic Ocean, as above alluded to, and which rises to such a

height that, according to the observations of Piazzi
Smyth on the Peak of Teneriffe, at a level of 10,700 feet
above the sea, it entirely obscures the sun before he sets
into the stratum of clouds below.   Such a current as this
must have a tendency to interfere with the free passage
of the upper or counter-trade-wind on its way to the
tropics, and to force it downwards into the lower or
direct Trade-wind, and the point where this intrusion
takes place must advance at the same rate as the upper
cross-current, which produces it, moves forward from E.
to W.   The interference of a current flowing from E. to
W., with another which is flowing from SW. to NE., must
necessarily generate a rotatory motion in a direction
opposite to that of the hands of a watch.   According to
this view, the cyclone which advances from SE. to NW.,
in the under Trade-wind, represents the advancing point
of contact of two other currents in the upper strata which
are moving in directions at right angles to each other.
This is the primary cause of the rotatory motion, and the
further progress of the cyclone will necessarily follow
according to the principles already laid down.   The
cyclone, being considered in this sense as the result of
interference of the currents at different points successively,
may thus preserve its diameter unchanged for a considerable period, or may even contract in dimensions, although
the case in which dilatation takes place will be more
usual.   The West Indian Islands are the boundary district
of two systems of meteorological conditions which offer
a strong contrast to each other ; characterised respectively
by the extent to which the periodical variation of pressure
takes place in the one, and by the absence of this variation in the other ; and are therefore especially liable to
the ravages of cyclones.

It is, moreover, perfectly clear that, if the above expla-

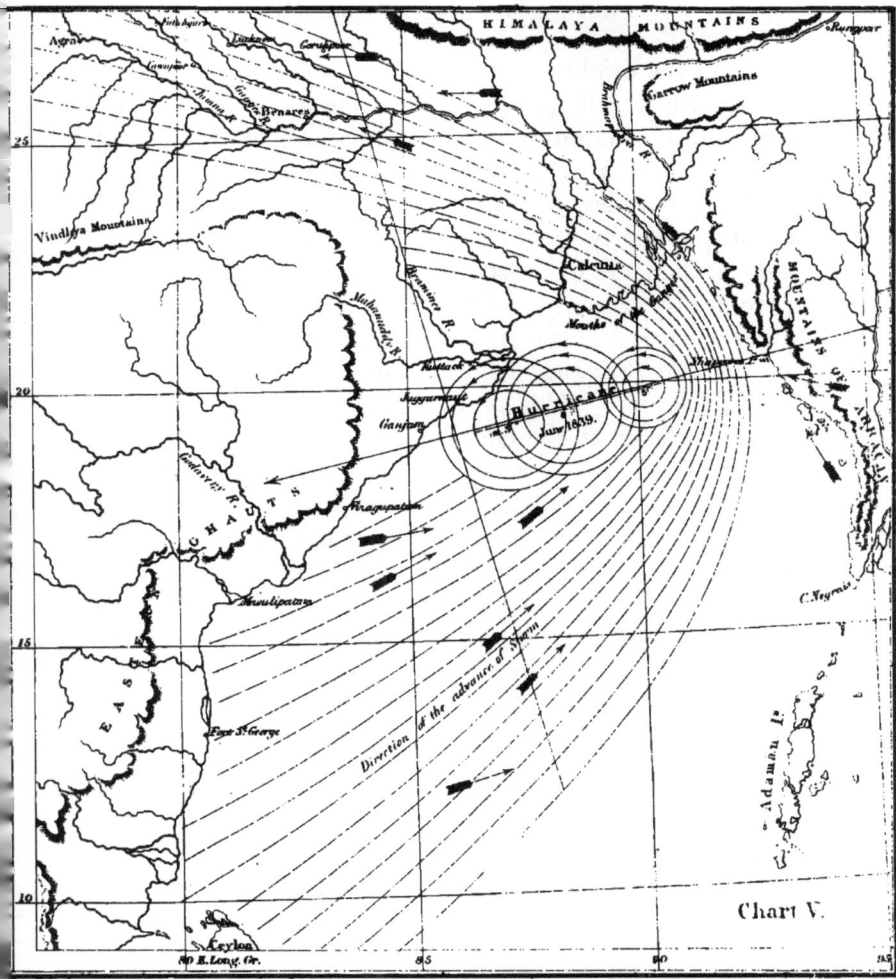

Chart V.

nation of the origin of the cyclonical motion be correct, a cyclone which will rotate in the same direction must be generated by the interposition of any mechanical obstacle in the path of a current which is moving towards high northern latitudes, which causes such a current to assume a more southerly direction (that of a south wind) on its eastern than on its western side, where it still remains more nearly westerly. This was the case with the storm which occurred in the Bay of Bengal on the 3rd, 4th, and 5th of June 1839, according to the observations collected by Piddington.* This storm was one of those which are not uncommon at the change from the north-east to the south-west Monsoon, which usually takes place between the 15th of May and the 15th of June in the Bay of Bengal.† It was, during the greater part of its course, a wind blowing and advancing in a definite direction (*a gale*, i. e. *a strong wind blowing in with tolerable steadiness from one quarter of the compass*), and at one point of its course became a cyclone (*a hurricane*, i. e. *a violent wind blowing in a circle or vortex of greater or less diameter*).‡ It blew (see Chart V.) as a violent SW.

---

* 'Researches on the Gale and Hurricane in the Bay of Bengal on the 3rd, 4th, and 5th of June, 1839, being a First Memoir with reference to the Theory of Storms in India.' Journal of the Asiatic Society of Bengal, No. 91, p. 550. Second Part, No. 92, p. 631.

† According to Brown's observations, the commencement of the south-west Monsoon took place at Anjarakandy, on the Malabar coast, from the years 1820 to 1833 inclusive, on the 20th, 31st, 31st, 27th May, 15th June, 21st May, 18th June, 26th May, 5th June, 9th and 26th May, 16th, 2nd, and 6th June; whilst, on the other hand, it set in at Canton, according to the 'Canton Register,' in 1830, from the 20th to the 28th April; in 1831, from the 7th to the 17th April; in 1832, from the 4th to the 7th April; in 1833, from the 9th to the 14th April; in 1834, from the 3rd April to the 8th May; and in 1835, from the 8th to the 21st April.

‡ This appears also to be the case with those storms which set in when the south-west gives place to the north-east Monsoon. These storms, which the Spaniards in the Philippines call ' *los temporales*,' are not accompanied by rain. The air is then obscured only by the thickly-drifting spray. On

Monsoon from the east coast of Ceylon and Masulipatam across the Bay of Bengal as far as the mountains of Arracan, where it turned at right angles and blew as a SE. storm over Calcutta and Benares to Lucknow, Cawnpore, and Agra in the Valley of the Ganges. The cyclone arose at the point where the storm turned, close to Arracan, in the focus of its parabolic course, as Piddington expresses it; it travelled in a direction from between ENE. and E. towards WSW. or W., past the mouths of the Ganges, from the Shapooree islands to Vizagapatam, Ganjam, Juggernaut, and the mouths of the Mahanuddy and Bramnee, and rotated, like the West India hurricanes in the direction S., E., N., W.

This proves that cyclones rotating in the same direction as those of the West Indies are produced, under conditions essentially different, by the mere fact of the motion of the mass of air becoming more southerly on its eastern side than on its western where it still remains westerly. It is possible that the typhoons of the China Seas may owe their origin to similar causes. The collective phenomena of the SW. Monsoon present us with analogous conditions. The direction of that current, which is SW. in the Indian Ocean and the Bay of Bengal, becomes more and more southerly in the China Seas, and it can only be decided by more extended observations in that district whether this deflection is due to the chain of the Philippine Islands, or is an immediate consequence of the contact of the Monsoons and Trade-wind. Horsburgh* says expressly, that along the south coast of China, the typhoons between July and September produce a rotation

---

the Coromandel coast these storms are called 'the breaking up of the monsoon.' On the Malabar coast the Portuguese distinguish those which are particularly violent by the name of *Elephanta*.

* India Directory, ii. p. 233.

of the vane from NW., through N., NE., E., SE., S., close to the coast, and further out to sea one from N., through NW., W., SW., S. This implies, in other words, that they are cyclones which rotate in the direction S., E., N., W., and travel along the coast, so that the coast feels the northern side of the storm, while, at a distance from the coast, the southern side is alone experienced. A recent instance of such a storm as this is given by the Raleigh Typhoon of August 5, 1835, which travelled from the Bashee Islands, between Luzon and Formosa towards Macao, in the direction from ESE. towards WNW., and corresponded in every particular to the description given.

If the cyclones be due to the fact that the direction of the SW. Monsoon becomes more southerly on its eastern side than on its western, and if, on this account, they travel from E. towards W., they ought to be especially prevalent in the eastern portion of the Indian Ocean. Dampier actually remarks that storms are expected on the Coromandel coast in April and September, the months of change of the Monsoon, but that they are more frequent on the Malabar course during the whole of the west Monsoon.

As regards the track of the typhoons, the following general charts have been published on the subject, viz. ' Piddington. The Southern Indian Ocean, with the average Tracks of its Cyclones, mostly from Colonel Reid, Mr. Thom, and Mr. Bosquet. 1850.'—' The Bay of Bengal, and Part of the Arabian Sea, with the Courses of their Cyclones from 1800 to 1848.'—' The China and Loo Choo Seas, and adjacent Pacific Ocean, with the Courses of their Cyclones from 1780 to 1847.'

These charts are also to be found in Piddington's work : ' The Sailor's Hornbook for the Law of Storms, being a practical Exposition of the Theory of the Law of Storms,

and its Uses to Mariners of all Classes in all Parts of the
World : shown by transparent Storm-Cards and useful
Lessons.   London, 8vo. 3rd ed. 1860, 408 pp.   Williams
and Norgate.'

Piddington has, if I mistake not, published thirty separate
papers on individual storms.   The annexed chart (VI.)
will be sufficient for our purpose.   The hurricanes of the
southern hemisphere are specially handled by Thom in
' An Inquiry into the Nature and Course of Storms in the
Indian Ocean, South of the Equator, for the practical
Purpose of enabling Ships to ascertain the Proximity
and relative Position of Hurricanes, with Suggestions
on the Means of avoiding them.   London, 1845, 8vo.
352 pp.'

From an examination of these instances the following
unmistakable conclusions may be drawn : —

(1.) That the rotation of the cyclone takes place in the
direction of the hands of a watch, which is opposite to
that which is observed in the northern hemisphere.

(2.) That the direction of its path within the torrid
zone is from NE. towards SW., and that it turns when
it reaches the Tropic of Capricorn, and advances from
thence in the direction from NW. towards SE.

(3.) That the level of the barometer sinks as the
observer approaches the centre, exactly as in the case of
the West India hurricanes.

A very good example of this is afforded by the storm
which travelled along the Malabar coast in April 1847,
with a velocity of advance of twelve miles per hour.   The
following tables (p. 193) give the heights of the barometer in its vicinity on the 17th and 18th of the month.
During this storm on the 18th, ten inches of rain fell on
Dodabetta, and twenty-nine inches at Tellicherry.

While the centre moved up along the Malabar coast

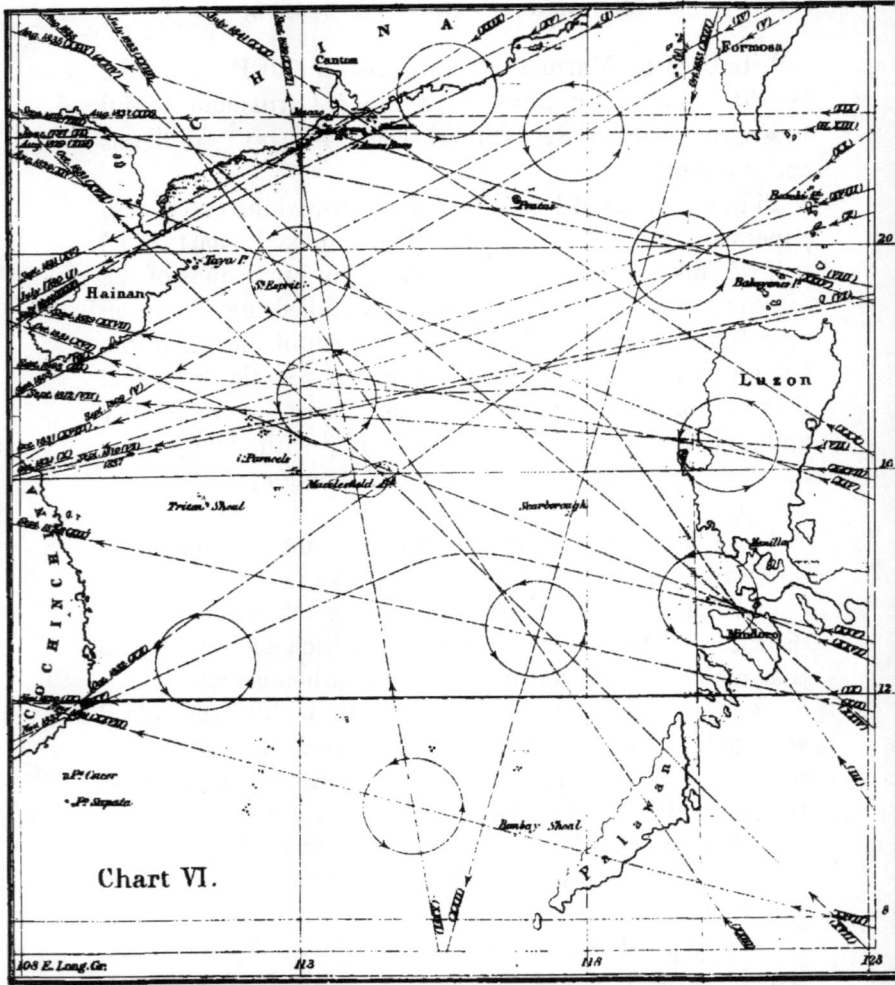

Chart VI.

Formosa

C H I N A

Canton

Macao

Hainan

COCHIN CHINA

Taya I.

S.t Esprit

i: Paracels

Triton. Shoal

Macclesfield B.k

Scarborough

Bashees Is.

Balingtang P.

Luzon

Manilla

Mindoro

Palawan

Bombay Shoal

n.P.t Cecer

P.t Sepata

20

16

12

8

108 E. Long. Gr.   113   118   123

the direction of the wind in the Arabian Sea was northerly, in the neighbourhood of Ceylon and at the equator westerly, in the Gulf of Bengal southerly. and south-easterly, in Scinde and Upper Bengal easterly.   In this case the tangential direction of the air which is in motion is very clearly marked.

APRIL 17.

	Distance from Centre Miles	Barometer Inches
Ship *Zemindar* (Lat. 6° 57' N. Long. 57° 12' E.)	800	30·00
Madras	°00	29·97
Colombo	300	29·86
Ship *Victoria* . . .	120	29·75
Cannanore. . . .	120	29·64
Ship *Buckinghamshire* .	90	29·58
„ *Mermaid* . . .	60	29·35

APRIL 18.

	Distance from Centre Miles	Barometer Inches
Ship *Zemindar* (Lat. 7° 42' N. Long. 57° 37' E.	900	30·00
Colombo	500	29·86
Madras	480	29·94
Rajcote	450	29·85
Bombay	240	29·70
Cannanore. . . .	130	29·78
Ship *Victoria* . . .	130	29·70
„ *Mermaid* . . .	80	29·36
„ *Buckinghamshire* .	20	28·35
Centre	0	28·00

Dampier, that prince of navigators, as the Englishmen with pride call him, gives the signs of an approaching Typhoon with that accuracy for which he is so remarkable. In his 'Voyages' (ii. 36) he says —

Typhoons are a particular kind of violent storms blowing on the coast of Tonquin and the neighbouring coasts in the months

of July, August, and September. They commonly happen near
the fall or change of the moon, and are usually preceded by
very fair weather, small winds, and a clear sky. Those small
winds veer from the common Trade of that time of the year,
which is here at SW., and shuffles about to the N. and NE.
Before the storm comes there appears a boding cloud in the
NE., which is very black near the horizon, but towards the
upper edge it looks of a dark copper colour, and higher still it
is brighter, and afterwards it fades to a whitish glaring colour,
at the very edge of the cloud. This cloud appears very amazing
and ghastly, and is sometimes seen twelve hours before the
storm comes. When that cloud begins to move apace, you
may expect the wind presently. It comes on fierce, and blows
very violently at NE., twelve hours more or less. It is also
commonly accompanied with terrible claps of thunder, large
and frequent flashes of lightning, and excessive hard rain.
When the wind begins to abate, it dies away suddenly, and,
falling flat calm, it continues so an hour, more or less; then
the wind comes about to the SW., and it blows and rains as
fierce from thence as it did before at the NE., and as long.

Varenius also gives a very accurate description of them
in his 'Geographia Naturalis,' book i. chap. xxi. § 12, as
follows :—

Violentus erumpens et rapida vertigine circa horizontem ro-
tatus, assiduis incrementis horarum circiter viginti spatio cir-
culum conficit, impetu horribili sævisque turbinibus vastè illa
æquora vehementissimè commovens. Autumnio maxime tem-
pore furiosissimus typhon dominatur sæpe tanto impetu et
rabie, quantam qui oculis non subjecerint, animo concipere non
possunt, dicas, cælum et terram chaos velle repetere. Neque
tantum in mari, sed in littoribus quoque furit multasque domos
prosternit, ingentes arbores radiciter evellit, magnas naves ex
mari in terram ad quadrantem milliaris propellit. Nautæ
vocant, ' de windt drayt rond om het compas.'

(It bursts forth with violence and shifts round the horizon
with a rapid rotating motion, increasing steadily, and completes

the circuit in the space of about twenty hours, producing the most vehement commotions in those seas by its raging fury and fierce whirlwinds. The Typhoon is most violent in autumn, and rages with such intensity and fury that those who have never seen it can form no conception thereof; you would say that heaven and earth wished to return to their original chaos. Its ravages are felt, not only at sea, but also on land, where it levels many houses with the ground, tears up huge trees by the roots, and propels great ships out of the sea to a distance of a quarter of a mile on the dry land. Sailors say ' the wind shifts round the compass.')

As a contrast to these descriptions, which were published two centuries ago, we find in Lecoq's 'Meteorology,' which appeared in 1836, at p. 467, the following explanation given by Page : —

Que sont les effrayans ouragans de la Mer de Chine, connus sous le nom de Lyfoongo, sinon des trombes, et des trombes immenses, terribles, qui, en crevant, se terminent par un violent coup de vent. (What else are the terrible hurricanes of the China Seas, known under the name of Lyfoongo (sic) but whirlwinds (trombes), and whirlwinds of a terrible and gigantic size, which, in bursting, end in a violent squall of wind.) Accordingly, a Typhoon is a tromb which bursts !

Having in this way discovered in the Typhoons a confirmation of the principle laid down for the origin of the rotatory motion in the case of the West India hurricanes, we may proceed to examine more closely the phenomena which accompany these great disturbances of the atmosphere.

If the rotating cylinder in a storm in the Trade-wind zone extends from the lower to the upper current, it is evident that the upper portion of the cylinder is exposed, in the counter Trade-wind, whose direction is southwesterly, to the same conditions which are presented to

the lower portion on its exit from the Trade-wind zone.
This upper part of the cyclone will accordingly dilate at
once, and advance in a direction different from that of
the other part. Hence, as a secondary phenomenon, a
suction will ensue in the centre of the cyclone, and also a
diminution of pressure on the surface of the earth. The
reason of this is twofold : the rotation of the cyclone
has a tendency to remove the air from the axis ; and the
cyclone expands like a funnel in its upper portion, so
that its upper strata have a greater centrifugal tendency
than the lower ones, to which, in consequence, there is
imparted a tendency to rise, in order to compensate for
the rarefaction which takes place above.

If we examine the observations closely, we shall find
that the storm itself does not owe its origin to any such
suction. As an instance, we shall take the hurricane of
August 2, 1837, for which we have simultaneous meteorological observations from St. Thomas and Porto Rico,
which I append in parallel columns, with a view to
comparison.

In the observations at St. Thomas the ' dead calm ' is
mentioned, which is suddenly interposed between the
two storms from the opposite quarters — an awful
moment, which fills the heart of the most fearless seaman with anxious suspense. Its occurrence is easily
explained on the hypothesis of a rotatory motion, as the
air at the centre of a cyclone must be at rest. It appears
irreconcilable with a centripetal motion of the air, since
two winds blowing directly against each other must check
each other's motion ; so that the intensity of both must
decrease the nearer the observer is to the point of contact of the currents. This actually happens on a great
scale with the Trade-winds, and must necessarily be
observable in the centre of a hurricane when it passes

over the point of observation, if a centripetal motion existed in them.   The observed facts exhibit phenomena the opposite of these.   According to Hoskiaer's observations at St. Thomas, the hurricane increased continually from 6^h 45^m to 7^h 30^m; then followed the dead calm, and at 8^h 10^m the hurricane recommenced as suddenly as it had ceased before.   How can anyone dream of two

St. Thomas				Porto Rico	
Mean Time		Barometer	Direction of the Wind	Barometer	Direction of the Wind
	h. m.	Inches		Inches	
August 1	18 0	29·932			
„ 2	2 10	·754	NW. ⎫		
	3 20	·666	N. ⎪ Gale fresh-		
	3 45	·666	N. ⎬ ening		
	4 45	·489	N. ⎪		
	5 40	·443	NE. ⎪		
	5 45	·310	NE. ⎭		
	6 30	29·133	NW. ⎫		
	6 35	28·911	NW. ⎪		
	6 45	·778	NW. ⎪		
	7 0	·778	NW. ⎬ Hurricane		
	7 10	·600	NW. ⎪		
	7 22	·289	NW. ⎪		
	7 30	·156	NW. ⎭		
	7 35	·111	⎫		
	7 52	·067	⎬ Dead Calm	8) 29·602	NNE.
	8 10	·067	⎪		
	8 20	·067	⎭		
	8 23	·423	SSE. ⎫		
	8 33	·511	SE. ⎪		
	8 38	·600	SE. ⎪		
	8 45	·689	SE. ⎪		
	8 50	·778	SE. ⎪		
	9 0	28·955	SE. ⎪	29·503	
	9 10	29·133	SE. ⎪		
	9 25	·222	SE. ⎬ Hurricane		
	9 35	·310	SE. ⎪		
	9 50	·399	SE. ⎪		
	10 10	·489	SE. ⎪	10) 29·402	
	10 35	·577	SE. ⎪		
	11 10	·599	SE. ⎪	11) 29·301	E.
	11 30	·621	SE. ⎪	12) 28·002	Hurricane
	14 45	·754	SE. ⎭	15½) 29·171	S.
	20 0	·888	SW.	16) 29·503	
	21 0	29·910	E.		

winds opposing each other in this instance ?   In addition
to this, the centripetal theory would require that the wind
at this time should have blown from Porto Rico to St.
Thomas, so that the wind at Porto Rico would have been
W. ; it was really NNE., exactly as would be produced
by the motion of a cyclone whose centre was at St.
Thomas.

It may also be shown that the diminution of barometrical pressure is not the cause of the violent disturbance of the air, but rather a secondary effect of it.
This appears from Hoskiaer's remark that at each gust
of the hurricane the level of the barometer sank 0·2
inches, but returned to its former height as soon as the
gust was over.   The same fact is observed in the
storms of the South Indian Ocean ; for Thom says
(p. 184) : —

It has been observed that the mercury in the tube is, in
certain parts of the storm, subject to sudden oscillations within
the space of a few minutes.

The sudden cessation of the storm appears as a normal
phenomenon in the following account. Nassau (Bahamas),
Aug. 1, 1813 : —

About half-past 2 in the afternoon the storm reached its
greatest height, and its fury lasted until 5, when it suddenly
ceased.   For half an hour afterwards it was succeeded by so
perfect a calm that it could only be compared with death after
the most frightful convulsions.   The inhabitants of the colony,
being well versed in the nature of storms, had recourse during
this interval to all possible precautionary measures, in order to
secure themselves against the second portion of the storm,
which they expected would make its appearance from SW., and
in fact it did burst forth again with the greatest fury about 6
o'clock.

The sudden transition from a violent storm to a dead calm at the centre, and then again to a storm whose direction is diametrically opposite to that which preceded it, is as well marked in the case of the Typhoons as in that of the West India hurricanes, as Dampier remarks. In the chapter on storms, in his ' Theory of the Winds ' (p. 69), he describes a storm which Captain Gadbury experienced at Antigua in August, 1681, in the following words : —

Before 8 o'clock the wind came on very fierce at NE., and, veering about to the N. and NW., settled there, bringing with it very violent rains.   Thus it continued about four hours, and then fell flat calm, and the rain ceased.

After this lull the wind and rain came on with the same violence as before, but from the SW.

In the beginning of the night, when the NE. gust raged, the sea ebbed so prodigiously, or else was driven off the shore by the violence of the wind so far, that some ships riding in the harbour in three or four fathom water, were aground, and lay so till the SW. gust came, and then the sea came rolling in again with such prodigious fury, that it not only set them afloat, but dashed many of them on the shore.   One of them was carried up a great way into the woods ; another was strangely hurled on two rocks that stood close by one another, with her head resting on one rock and her stern on the other ; and thus she lay like a bridge between the two rocks, about ten or eleven feet above the sea, even in the highest tides : for the tides do usually rise here but little, not above two or three feet, but in these hurricanes it always ebbs and flows again prodigiously.*

* In the description of the storm which devastated Charleston on September 15, 1752, we read : ' All the vessels in the harbour driven ashore, and some of them six miles inland, over the marshes and small streams, the inhabitants taking refuge in the upper parts of the houses, as in each previous case.'

It was not the ships only that felt the fury of this storm, but the whole island suffered by it; for the houses were blown down, the trees tore up by the roots, or had their heads and limbs sadly shattered, neither was there any leaves, · herbs, or green thing left on the island, but all looked like winter; insomuch that a ship coming thither a little after, that used that trade, could scarce believe it to be the same island.   . . .

.   .   The day after the storm the shore was strewed with fish of divers sorts, as well great as small; such as porpoises, sharks, &c.; and abundance of sea-fowl also were destroyed by it. . . . .

Though I have never been in any hurricane in the West Indies, yet I have seen the very image of them in the East Indies, and the effects have been the very same: and for my part I know no difference between a hurricane among the Carribee Islands in the West Indies, and a Tuffoon on the coast of China in the East Indies, but only the name. I have given a large account of one of these in my ' Voyage round the World,' chapter xv. p. 414.

This Typhoon came first from the NE., then there was a calm, then a storm from the SW., as violent as the previous one. During the storm a *corpus sant* appeared at the mast-head.

At Madras, on October 13, 1836, between 6 A.M. and 4 P.M., the barometer fell, with a wind from the N., from 29·940 inches to 29·111; and from that to 7 P.M., to 28·285, during the hurricane from the north. From 7^h 15^m to 7^h 45^m there was an awful calm, the barometer remaining at the same level, 28·285 inches, for half an hour. The hurricane from the south then burst forth, and the barometer rose to 29·001 inches before 9 o'clock, and to 29·415 before 11.

During the Mauritius storm of March 1836 (p. 160), the calm on the 6th lasted two hours; the fall of the barometer previous to it was 1·511 inches in thirty-three hours, with easterly winds ; its subsequent rise lasted for thirty-seven hours, with westerly winds (according to the 'Nautical Magazine' for June 1837).

The circumstances of the storm of October 24, 1818, were exactly similar to those described, as we read in Goldingham's account : —

In the morning the wind blew strong from N., and before 10 o'clock it freshened to a gale. Then succeeded a frightful lull for half an hour, after which it blew a perfect hurricane from S. with a fury which had never been known in Madras before. The barometer had fallen to 29·5 inches, but during the fearful calm at 10 o'clock it stood at 28·78 inches, an unusual and alarmingly low level, of which I had never before heard at that place : towards noon it had risen about half an inch.

The directions of the wind at the cessation and recommencement of the storm represent the tangents of the innermost circle, so that, whenever the path of the cyclone coincides with a diameter of the circle, these tangents will be parallel, and consequently the directions of the wind diametrically opposite to each other. Whenever the path coincides with a chord, the directions will be inclined to each other, and the duration of the lull at the centre be proportionably diminished. A very good example of this is given by the Typhoon which Krusenstern experienced in the Japanese Seas, in lat. 31° N., long. 128° W. The directions of the wind before and after the lull were ESE. and WSW. The critical moment at which the change took place was preceded by a complete calm, which only lasted for a few minutes, and during which the barometer fell to 27 inches. The calm was not of longer duration than the above in the case of the hurricane which devastated New Orleans in 1776, as we see from the description by Dunbar, in the 'Trans. of the

American Philosophical Society,' vol. vi.   Thom gives the
duration of the calm in the *Albion* storm, as 1 or 2 hours;
in that of the *Mauritius*, in 1836, as 3 or 4 hours at Port
Louis (a longer period than is given by the Journal of the
Observatory); in the *Carnatic* storm 4 hours; in the
Neptune storm 6 hours; and in the *Windsor* storm 8
hours.   Thom gives the diameter of the calm area at 80
miles in the *Windsor* storm; in the case of the *Rodriguez*
storm at 30 or 40 miles at first, at 20 subsequently in lat.
22° S.   Dr. Finlay, a physician in Havannah, has communicated to me the interesting fact, that when Havannah
was ravaged by a frightful hurricane on October 12,
1846, the fall of the barometer at the centre of the
cyclone was so sudden that the windows of the houses
were blown *outwards*.[*]

We have as yet, in speaking of the advance of the
cyclone, taken no notice of the resistance offered by the
ground to the motion of the air.   This resistance will, as
Redfield correctly remarks, cause the rotating cylinder to
incline forwards in the direction of its course.   In consequence of this the cyclone will commence in the upper
strata before it does so below, so that the barometer will
begin to fall before the storm commences, and will indicate its approach.   The oblique direction of the axis will
cause the cold upper strata to intermingle with the warmer
strata below, so that copious precipitations of moisture will

[*] 'On this account, windows are sometimes set open at the beginning of
a hurricane.' (Note by Adm. Fitzroy to the translation of the first edition.)
Similarly, Tooley says of the tornado which devastated Natchez in 1840:
'The destruction was produced by the expansion of the air shut up in
houses. Of all which stood in the path of the hurricane, those only escaped
which had their doors and windows open.' Results directly opposite to
these were observed at Tuscaloosa in Alabama during the tornado of March 4,
1842. In this case, the only houses that were destroyed were those whose
doors and windows were open, while those which were quite shut up were
preserved.

result, and will be accompanied by electrical explosions, which will be the more violent the heavier the fall of rain is. Thus the cold air will appear to pour down out of the clouds, and the storm assume the form so aptly termed by the Greeks ἐκνεφίας.* With this might be classed the phenomenon well known to sailors in the torrid zone, under the name of the 'Bull's eye,'† the small black cloud which suddenly appears in rapid motion in the sky, seems to grow out of itself, and soon covers the whole sky, producing a disturbance of the atmosphere, which is the more frightful the finer the weather was which immediately preceded it.

Hitherto our investigations have been confined to the consideration of the occurrences which take place within the cyclone itself, and not of the phenomena observed at the places over which it passes. If we collect these places together in our minds, and consider, in order to simplify matters, that the diameter of the cyclone remains unaltered during its advance, we shall find that the points which it merely touches lie on two lines, which are both of them parallel to the path of the centre, as they are formed by the common tangents of a system of equal circles. If the cyclone enters the temperate zone, and advances from SW. to NE., the direction of the wind on the upper line of contact will be NE., on the lower SW. Along the former line the NE. wind will commence in the south before it does so in the north, and thus seem to work back, while along the lower line the SW. wind will appear to advance in the direction in which it blows.

* 'A hurricane, caused by the meeting and bursting of clouds.' Liddell and Scott.

† This should not be confused with the 'eye of the storm' (p. 282), which is an opening between the clouds. The 'bull's-eye' is the cloud referred to by Sir J. F. W. Herschel (Encyc. Brit., article 'Meteorology '), F.
Vide also 1 Kings xviii. 44. Trans.

Franklin writes on May 12, 1760, in a letter to Alexander Small —

I think that the NE. storms in North America commence first, as regards time, in the southern part of the country : i. e. that the air in Georgia, the most southerly of the colonies, begins to move to the south-west sooner than the air in Carolina, which is the next colony to it on the north-east side; that in Carolina sooner than that in Virginia, and so on through Pennsylvania, New York, and New England, to Newfoundland.

Franklin made this discovery on the occasion of an eclipse of the moon, which he was unable to observe at Philadelphia, owing to a NE. storm; while at Boston, where the storm began later, the eclipse was observed. According to the investigations of Bache this eclipse took place on October 21, 1743.

Lewis Evans says the same, on a map of Pennsylvania, published in 1747. 'All our great storms begin to leeward, thus a NE. storm shall be a day sooner in Virginia than in Boston.'

Mitchel found in the year 1802 that a NE. storm commenced

At Charleston, Lat. 34° 45'	on February 21,	at	2 o'clock P.M.
Washington „	38	55	„	„	5 „
New York „	40	40	„	„	10 „
Albany „	44	0	„	22	Daybreak.

In the year 1811

At Cape Hatteras, Lat. 34° 15'	on December 23,	at	8 o'clock P.M.
Washington „	38	55	„	„	10 „
New York „	40	40	„	„	Midnight.
Lyme			„	24	2 o'clock A.M.
Boston „	42	22	„	„	4 „

Poev alludes to a cyclone which travelled from Charleston

to Nova Scotia, between February 21 and 23, 1802. I can find no accounts of the other two cases. They may be simply explained on the hypothesis of a cyclone sweeping along the coasts of the United States, but might also belong to a class of storms, which we shall describe subsequently when we treat of the temperate zone.

In order to be enabled to give a more correct description of the phenomena which accompany West India hurricanes, we borrow from Major-General Reid's work an account of the Barbadoes hurricane of August 10, 1831, as, according to that author, it was obtained from an eye-witness :—

At 7 P.M. the sky was clear and the air calm ; tranquillity reigned until a little after 9, when the wind blew again from the north ; distant lightning was observed at half-past 10 in the north-north-east and north-west. Squalls of wind and rain from the north-north-east, with intermediate calms succeeding each other until midnight ; the thermometer fell to 83° Fahrenheit, and during the calms it rose as high as 86° ; after midnight the continued flashing of the lightning was awfully grand, and a gale blew fiercely from the north and north-east ; but at 1 A.M. on August 11 the tempestuous rage of the wind increased : the storm, which at one time blew from the north-east, suddenly shifted from that quarter, and burst from the north-west and intermediate points. The upper regions were from this time illuminated by incessant lightning, but the quivering sheet of blaze was surpassed in brilliancy by the darts of electric fire which were exploded in every direction.*

* If we remember that dust and sand, carried up mechanically into the air, especially when whirled about, can produce electrical tension, which may even rise to such a degree that bright sparks pass ; and, further, that it is the blackest clouds of smoke from a volcano which are the chief seat of the flashes of lightning, called by the inhabitants of the towns on Vesuvius *ferilli*, and which appeared in the same clouds at the elevation of the islands of Sabrina and Ferdinandea ; and, in addition to these instances, combine the fact of electrical appearances, being visible at waterfalls with

At a little after 2 the astounding roar of the hurricane, which rushed from the north-north-west and north-west, cannot be described by language. Lieutenant-Colonel Nickle, commanding the 36th Regiment, who had sought protection by getting under an arch of a lower window outside his house, did not hear the roof and upper story fall, and was only assured this had occurred by the dust caused by the falling ruins. About 3 the wind occasionally abated, but intervening gusts proceeded from the south-west, the west, and west-north-west, with accumulated fury.

The lightning also having ceased, for a few moments only at a time, the blackness in which the town was enveloped was inexpressibly awful; fiery meteors were presently seen falling from the heavens; one in particular, of a globular form and a deep red hue, was observed to descend perpendicularly from a vast height. It evidently fell by its specific gravity, and was not shot or propelled by any extraneous force: on approaching the earth with accelerated motion it assumed a dazzling whiteness and an elongated form, and, dashing to the ground in Beckwith Square, it splashed around in the same manner as melted metal would have done, and was instantly extinct. In shape and size it appeared much like a common barrel shade:* its brilliancy and the spattering of its particles on meeting the earth gave it the resemblance of a body of quicksilver of equal bulk. A few minutes after the appearance of this phenomenon, the deafening noise of the wind sank to a solemn murmur, or, more correctly expressed, a distant roar, and the lightning, which from midnight had flashed and darted forkedly with few and but momentary intermissions, now for a space of nearly

the frequently-observed phenomenon of flashes of light when rain-drops touch the ground, it seems not unlikely that, as Faraday's experiments with the hydro-electrical machine seem to show, the source of the electricity in a thunderstorm is to be found in the friction of the steam in its condensation into the form of water. The more considerable the condensation, and the more active the disturbance of the air, the greater will be the amount of electricity generated. Country people are quite right when they call a downpour of rain a 'silent thunderstorm.'

* A 'barrel shade' is the name for the glass cylinder put over candles in the tropics.

half a minute played frightfully between the clouds and the earth with novel and surprising action. The vast body of vapour appeared to touch the houses, and issued downward flaming blazes, which were nimbly returned from the earth upward.

The moment after this singular alternation of lightning, the hurricane again burst from the western points with violence prodigious beyond description, hurling before it thousands of missiles — the fragments of every unsheltered structure of human art. The strongest houses were caused to vibrate to their foundations, and the surface of the very earth trembled as the destroyer raged over it. No thunder was at any time distinctly heard. The horrible roar and yelling of the wind,*— the noise of the ocean, whose frightful waves threatened the town with the destruction of all that the other elements might spare, — the clattering of tiles, the falling of roofs and walls, and the combination of a thousand other sounds, formed a hideous and appalling din. No adequate idea of the sensations which then distracted and confounded the faculties can possibly be conveyed to those who were distant from the scene of terror.

After 5 o'clock the storm, now and then for a few moments abating, made clearly audible the falling of tiles and building materials, which by the last gust had probably been carried to a lofty height. At 6 A.M. the wind was at south, and at 7 south-east, at 8 east-south-east, and at 9 there was again clear weather.

As soon as dawn rendered outward objects visible the writer proceeded, but with difficulty, to the wharf. The rain at the time was driven with such force as to injure the skin, and was so thick as to prevent a view of any object much beyond the head of the pier. The prospect was majestic beyond description:

* The frightful noise at the centre of cyclones is very variously described. Thom says, 'An awful silence, followed by an awfully hollow and distant rumbling noise.' Biden says that the gusts which succeed it are 'like to successive and violent discharges of artillery, or the roaring of wild beasts.' Cattermole says, 'A continuous roar in the air;' while Piddington says that the usual expressions for waterspouts are 'rumbling' and 'hissing,' while for cyclones they are 'roaring,' thundering,' 'yelling,' and 'screaming.'

the gigantic waves rolling onwards, seemed as if they would defy all obstruction; yet as they broke over the careenage they seemed to be lost, the surface of it being entirely covered with floating wrecks of every description; it was an undulating body of lumber, shingles, staves, barrels, trusses of hay, and every kind of merchandise of a buoyant nature. Two vessels only were afloat within the pier; but numbers could be seen which had been capsized, or thrown on their beam-ends in shallow water.

On reaching the summit of the cathedral tower, to whichever point of the compass the eye was directed, a grand but distressing picture of ruin presented itself: the whole face of the country was laid waste; no sign of vegetation was apparent, except here and there small patches of a sickly green. The surface of the ground appeared as if fire had run through the land, scorching and burning up the productions of the earth; the few remaining trees, stripped of their boughs and foliage, wore a cold and wintry aspect; and the numerous seats in the environs of Bridgetown, formerly concealed amid thick groves, were now exposed and in ruins. From the direction in which the cocoa-nut and other trees were prostrated next to the earth, the first that fell must have been blown down by a north-north-east wind; but far the greater number were rooted up by the blast from the north-west.

Thus far goes the account from Barbadoes. If we add to this that when the storm was at its height the electrical tension of the atmosphere was so great that sparks leaped out of a negro in the garden of Codrington College, we shall perhaps be prepared to admit with General Reid that the cause that a great number of forest trees at St. Vincent were destroyed without being blown down, is to be sought in this excessive amount of free electricity. Another phenomenon which accompanied this hurricane, and has often been observed elsewhere, was salt rain. At the N. point of Barbadoes the waves broke constantly over a cliff more than 70 feet high. The salt water was carried from this for miles into the country, so that all the fresh-water fish in the ponds of Major Leacock were killed; and at Bright Hall, two miles SSE. of the point, the water in the ponds was salt for many days after the storm.

We shall close our description with the account of the hurricane of October 10, 1780 (the 'Great Hurricane'), which seems to have presented in itself all the terrors of these grand natural phenomena. We possess more detailed accounts of it than of any other, because it was experienced in the West Indies by the fleet of Admiral Rodney, which had already suffered considerably a week previously to it, from the hurricane of Savannah-la-Mar on the west coast of Jamaica. Of that fleet the *Scarborough, Barbadoes, Victor,* and *Phœnix,* which had been lying in Montego Bay, were then lost. A ship, the *Princess Royal,* which was lying at Savannah-la-Mar with the *Henry* and *Austin Hall,* where they were all blown from their anchors and carried into the marshes, was lifted so far on the dry land that she served as a habitation for the surviving inhabitants of the place.

The diameter of the hurricane of October 10 was from the very commencement so great that it embraced the extreme points of the Leeward Islands,— that is to say, Trinidad and Antigua, — while its centre advanced on the 10th over Barbadoes as far as St. Lucia, where Admiral Hotham lay with the *Vengeance, Montagu, Egremont, Ajax, Alcmena,* and *Amazon.* Subsequently it overtook, on the south coast of Martinique, the French convoy, which consisted of 50 merchant vessels and transports, with 5,000 troops on board, escorted by the frigates *Ceres* and *La Constante:* only six or seven of these vessels were saved. ' Les bâtiments du convoi disparurent' (the vessels of the convoy disappeared), was the rather laconic expres-

P

sion made use of in the report given by the intendant of Martinique. From thence the vortex of the hurricane passed over Porto Rico, where the *Deal Castle* was wrecked, to the island of Mona, and there on the morning of the 15th it overtook the English merchant fleet convoyed by the *Ulysses* and *Pomona*, to which it did considerable damage. After this it moved on to Silver Keys, where the *Stirling Castle* foundered. At what spot the *Thunderer*, which was on her passage from St. Lucia to Jamaica, bearing the broad pennant of Commodore Walsingham, was lost, has never been ascertained. It now turned to the NE., when in 26° N. lat., and overtook the vessels of Admiral Rowley's squadron, which had been dismasted by the hurricane at Savannah-la-Mar, and consisted of the *Trident*, *Ruby*, *Bristol*, *Hector*, and *Grafton*, which, very unfortunately, were steering from the west side of the storm straight towards its centre. From thence it turned towards the Bermudas, probably extending across the Atlantic Ocean at the time when its breadth was greatest; and here the *Berwick*, which was on her return to England, after having been disabled by the first storm, fell in with it. A sketch of the *Egmont*, taken by one of her officers, and to be found in Reid's work, gives a good idea of the condition of the vessels. But not less destructively did the hurricane rage on the islands themselves. At Martinique 9,000 people perished, 1,000 at St. Pierre alone, where not a house remained standing, as the sea rose to a height of 25 feet, and 150 houses along the beach disappeared in a moment. At Port Royal, the cathedral, seven churches, and 1,400 houses were thrown down, and 1,600 sick and wounded buried under the ruins of the hospital, so that but few were saved. At Dominica almost all the houses standing near the shore were carried away, and the royal bakehouse, the magazine, and a part of the barracks, destroyed.   At St. Eustatia seven vessels were dashed to pieces on the rocks at North Point, and of 19 vessels which cut their cables and went to sea but one returned.   At St. Lucia, where 6,000 people were deprived of life, the strongest buildings were shattered to their foundations ; cannon were carried more than a hundred feet, and men and animals were raised from the ground and hurled to a distance of several yards.   The sea rose to such a height that it demolished the fort, and hurled a vessel against the naval hospital, which was crushed by its weight.   The coral bed at the bottom of the sea was torn up, and pieces of it raised so high on their edges that they were afterwards visible above the water ; and the harbour itself was deepened six feet, and in some places even more.   Of 600 houses at Kingstown, in the island of St. Vincent, only fourteen remained standing ; and the French frigate *Juno* was wrecked.   Sir G. Rodney in his official report says —

It is impossible to describe the dreadful scene it has occasioned at Barbadoes, and the condition of the miserable inhabitants. Nothing but ocular demonstration could have convinced me that it was possible for the wind to cause so total a destruction of an island remarkable for its numerous and well-built habitations; and I am convinced that the violence of the wind must have prevented the inhabitants from feeling the earthquake, which certainly attended the storm.   Nothing but an earthquake could have occasioned the foundations of the strongest buildings to be rent ; and so total has been the devastation, that there is not one church, nor one house, as I am well informed, but what has been destroyed.

In the Leeward Islands, the family at Government House retreated to the centre of the building as the tempest increased, imagining from the prodigious strength of the walls, they being three feet thick, and from its circular form, it would have withstood the wind's utmost rage; however, by half-past 11 they were obliged to retreat to the cellar, the wind having forced its passage into every part, and torn off most of the roof. The water rose to a height of four feet in the cellar, so that they fled to the batteries and sought shelter under the cannon, many of which were moved. So violent was the storm, *when assisted by the sea*, that a twelve-pounder gun was carried from the south to the north battery, a distance of 140 yards.* When the day broke the landscape had a complete winter aspect: not a leaf, not a twig, was to be seen on the trees.

In such a conflict of the elements, the strife of man is arrested. When the *Laurel* and *Andromeda* were wrecked at Martinique, the Marquis de Bouillé sent the twenty-five English sailors, who were the only survivors of the crews, to the British Governor at St. Lucia, with a letter stating that he could not detain them as prisoners, from the chances of a catastrophe common to all.

We now turn to the storms on their entry into the temperate zone.

The change in the direction of advance of the storms, on crossing the limit of the Trade-wind zone, from a south-easterly to a south-westerly,† has been explained by the fact that in the later part of their course the prevalent direction of the wind is SW., while in its earlier portion NE. winds interfered with their motion. The direction of the wind in the temperate zone is not constant but variable. Hence, phenomena similar to those observed in the torrid zone can only occur in the temperate when

---

* On its carriage, of course, which had wheels.

† [i.e. they first move *from* the SE., and afterwards *from* the SW. — Trans.]

the prevalent direction of the wind in that zone is SW. before the entry of the storm into it. Accordingly barometrical minima, connected with storms, will only occur when these conditions are fulfilled. This was very remarkably the case with the minimum of December 24, 1821, which has been described above. In November and December the mean direction of the wind was SW. at Penzance, London, Bushey Heath, Cambridge, New Malton, Lancaster, Manchester, Paris, Brest, Dantzic, Königsberg, &c. ; and according to the 'Bibliotheca Universalis,' 19, 29, a south-west wind, of more or less intensity, prevailed over the central region of Western Europe.

We have already explained the reason why the diameter of the cyclone should increase, and its intensity decrease, on the change in the direction of its path.* On the other hand, it is evident that the intensity will increase afresh if smaller cyclones are developed, by any means, from the larger ones. These conditions existed in the case of the minimum of December 24 in the Mediterranean, in which district the mass of air, being checked in its progress by the mountains of Spain and the Maritime Alps, set itself afresh in violent motion round these points as new centres. This explains why the hurricane exhibited such remarkable violence in the region referred to, as well as near its centre. Hence we have from Brest on the 26th : ' For the last fortnight we have been living in storms which have never ceased roaring with unparalleled

---

* It is not possible to ascertain the ratio between the intensities of storms in the torrid and temperate zones, in the absence of observations carried on with anemometers of similar construction. The mean intensity, in miles per hour, found by the observations in Liverpool from 1852 to 1855, was — winter, 15·6; spring, 12·1; summer, 11·8; autumn, 11·5: year, 12·75. The maximum was 62 on January 9, 1852. The observations at Kew give 10·36 miles for 1856, and 9·76 for 1857, giving as mean result 10·06 miles per hour. In the year 1857 this mean result was, at Liverpool, 11·5; Kew, 9·76; Oxford, 9·76; — giving as a general average 10·34 miles per hour.

fury.' In London there were the greatest floods that had
been known since 1809. In Portsmouth there was such
a squall from the SSE. as had not been felt for years.
The sea rose to an enormous height during it. The destruction in the Mediterranean was not less extensive.
From Leghorn to Barcelona fearful damage was done.
On the southern slopes of the Alps torrents of rain fell
which inundated Venice, Genoa, and Nice. In Appenzell
there was a storm such as had never been known there
before, and the fury of the hurricane was especially intense in the valleys, inasmuch as the mountains checked
the progress of the wind in such a manner that the barometer on their southern side suddenly rose to a much
greater height than that at which it stood on their
northern slopes.

We see, accordingly, that barometrical minima of the
temperate zone, as results of cyclones which enter that
region, are distinguished from those of the torrid zone, on
the one hand, as regards the direction in which they advance, on the other as regards the greater area over
which they are felt. In the case of the minimum of
August 2, 1837, the difference of barometrical pressure
between St. Thomas and Porto Rico was 1·33 inches, the
distance being barely 100 miles, while on December 24,
1821, the difference between Brest and Bergen was only
1·066, the absolute minimum reading being the same in
the two hurricanes. On May 21, 1823, the barometer
fell, on board the *Duke of York* off the Hidgelee coast,
from 28·866 inches to below 26·469 inches between 8
and 11 o'clock A. M. ; a distance of 2·397 inches in three
hours, according to both barometer and sympiesometer;
that is to say, the fluid disappeared for half an hour in
the casing of both instruments, which covered the scales
up to that level. The fall observed in Calcutta at the same

time was only 0·710 inch.   Accordingly we see that in
the torrid zone the fall of the barometer which precedes
the minimum, as well as the subsequent rise, is much more
rapid than in the temperate zone.

If we look to the total diminution of pressure, we shall
find this more extensive in the temperate than in the
torrid zone.   The former may be compared to a wide
valley with gently sloping sides, the latter to a deep and
precipitous ravine.   There must therefore be some other
cause in the temperate zone, in addition to those in the
torrid, which assists in producing the rarefaction.   This
is presented by the high temperature which the air in its
rapid advance into high latitudes l rings with it.   This
was very remarkable on December 24, 1821.   At Tol-
mezzo the thermometer rose to 88°·25 in the shade.   At
Geneva it rose rapidly 11°·25 on the night of the 24th,
and stood at its highest level 60°·1 at 1·30 A.M. on the
morning of the 25th.   Similarly, an unusual degree of heat
was observed at Boulogne, Paris, and Hamburg.   It is clear
that if so warm a current flows over Europe towards the
pole, the displaced polar current must force its way to the
southward somewhere else.   From the rotation of the
cyclone we should expect to find this current in America,
and it is actually indicated by the observations.   On
December 24 the thermometer stood at 9°·1, and a few
days later at + 0°·1 at Salem, Massachusetts, in the latitude
of Rome.   All the accounts from America mention an
extraordinary degree of cold.

The occurrence of such phenomena as these is not
confined to the winter months.   The hurricane which
devastated St. Thomas and Porto Rico on August 2, 1837,
was followed in the middle of the month, and on the 21st,
by two very violent storms, which are fully described
in Reid's work.   During this period the heat in Europe

was unusually great, and violent thunderstorms with torrents of rain were experienced in various places. At Messina the temperature was 99°·5 from the 10th to the 20th of the month; at Naples it ranged from 95° to 99°·5; at Rome it was 99°·5 on the 12th. In the valleys of the Roth and the Emme the rivers, swollen by the torrents of rain (*Wolkenbrüche*), carried down masses of rock weighing 3 tons.* In Silesia and Mark Brandenburg the heat was very oppressive. However, in Gallicia and Prussia this extraordinary heat was followed towards the end of the month by remarkably cold weather, which had already been felt in America at the time of the great heat in Europe. At Rochester, in the State of New York, the unusual phenomenon of a hoar frost was experienced on August 4.

If we are prepared to recognise in these meteorological phenomena of the temperate zone the influence of disturbances in the tropical atmosphere following each other in rapid succession, we shall at once see why any deviation in the direction of change of the wind from that of S., W., N., E., which is conformable to the Law of Gyration, must be a sign of unsettled weather. This has been remarked by almost all observers who have subjected the connection between the direction of the wind and the conditions of the weather to an accurate examination. The gyration of the vane produced by a cyclone at points

---

* In the valley of the Emme there is an old legend preserved, to the effect that, in the ravines of the Hohgant, there slumbers a huge serpent, which for centuries never appears, but then suddenly bursts forth with uncontrollable fury. In this description it is easy to recognise the mountain torrent, which, when swollen by the rain, bursts forth from the windings of the valley. Nothing had been known of it within the memory of man (since 1764), until it broke forth in August 1837. The pretty tale, *Die Wassersnoth in Emmethale vom August* 13, 1837, by the author of the *Bauernspiegel*, contains a most complete account of this great natural catastrophe.

situated on its north-west side is in the direction NW., W., SW., while in the ordinary course of things it should be in the opposite direction, viz. SW., W., NW.

From this circumstance Lloyd[*] has drawn the conclusion that cyclones occur much more frequently in Ireland than had been previously supposed. Martin, in a 'Memoir on the Equinoctial Storms of March—April, 1850; an Enquiry into the Extent to which the Rotatory Theory may be applied,' 1852, has given a Summary of Storm Tracks for the North Atlantic Ocean for the period assigned.

Redfield has done the same for the northern portion of the Pacific Ocean, in his ' Observations in relation to the Cyclones of the Western Pacific.' In the chart which he gives, we only find one cyclone whose advance in the torrid zone from SE. to NW., and in the temperate from SW. to NE., has been accurately traced, viz. the cyclone of the Mississippi between October 3 and 8, 1854. In all the other cases he was unable to obtain proofs of more than one of these motions ; either that from SE. to NW. in the torrid zone, or that from SW. to NE. in the temperate. The fact that they commence on the coast of California as SE. winds agrees with the hypothesis that they are cyclones, as such a direction of the wind would be

---

[*] 'Notes on the Meteorology of Ireland, deduced from the Observations made in the year 1851, under the direction of the Royal Irish Academy,' by Rev. H. Lloyd, D.D. Trans. R. I. A. xxii. 411. This author determines the position of the vortex at any instant, as follows: ' If y and x denote the distances (in geographical miles) of the place of observation from any assumed central point, measured on the meridian, and on the perpendicular to the meridian respectively, $y_{,}$ and $x_{,,}$ the corresponding co-ordinates of the centre of the vortex, and θ the angle which the direction of the wind at the point (y, x) makes with the meridian, measuring from N. to E. ;

$$y - y_{,} + (x - x_{,}) \tan \theta = 0.$$

The direction of the wind being perpendicular to the line connecting the points (y, x) and $(y_{,}, x_{,})$, $y_{,}$ and $x_{,,}$ are determined by combining the equations of the preceding form for each of the stations, by the method of least squares.'

that observed in front of a cyclone which is advancing in a direction perpendicular to the coast line. As for the Typhoons which strike the coast of China, and approach it from the SW., the point where they change their course must be on the continent, and their subsequent progress in the temperate zone could be only traced in the Japanese Seas. These Typhoons are frequently accompanied by extensive inundations of the coasts by the sea, of which Macgowan's work* contains numerous instances taken from Chinese sources. In the year 1748, 20,000 persons were destroyed by an inundation of the sea during a storm of this character. The violence which these storms exhibit even on the eastern edge of the Monsoon region is apparent from the fact, that in one storm which passed over the island Guaru,one of the Ladrone Islands, in September 1855, more than 8,000 people were rendered houseless in the space of twenty minutes by the fall of their roofs.

It is moreover evident that the connection between a storm in the temperate zone, and the original cyclone in the torrid zone, to which it owes its origin, need not necessarily be traceable, as a continuously advancing minimum, in the lower strata of the atmosphere. We have explained above (p. 36) the circumstance that the predominant winds in the upper strata, between the topics, are SW., so that the identical conditions are from the very first presented to the upper part of the cyclone, which cannot be experienced by the lower part before it leaves the zone of the Trade-winds. Hence the upper portion will dilate at once, and advance in a direction different to that in which the lower portion moves. A very good proof of this connection, which was pointed out by me

* On the Cosmical Phenomena observed in the neighbourhood of Shanghai during the past Thirteen Centuries.—*Journal of the North China Branch of the Royal Asiatic Society*, 1860, ii. 55.

in the year 1842, is given by the hurricane at the Havannah on October 12, 1846, in which case the effect of the diminution of pressure has been already alluded to (p. 202),* and the hurricane of October 17, in the same year in the south of France, which was accompanied by torrents of rain and the fall of dust. The description given by Lortet of this storm reminds us at once of the description of a tropical cyclone.

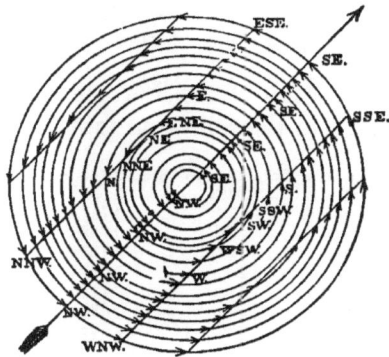

The tempest at La Verpellière, between Lyons and Grenoble, came over the chain of hills in the Ardeche district with a NW. wind. At the same time, from seven o'clock on the morning of October 17, the sky became extraordinarily overcast over Grenoble. The stifling blasts of a south-easterly sirocco were felt there; blood-coloured rain fell there, together with a reddish dust, which covered the diligences near Lyons to the depth of one or two lines. This occurred, according to Dupasquier, during a calm, or while the wind was from the S., and very light. The rain was not excessive; but the appearance of the sky was terrific. Two banks of clouds — one in the S., the other in the NW. — were the foci of the storm. The wind shifted from minute to minute; flashes of lightning of extraor-

---

* The fall of the barometer, for each hour, if we count the time from the period at which the centre reached the town, is as follows (in inches) : —

Between 12 and 6 hours before,	0·077	Between 4 and 2 hours before	0·290
„ 9 6 „ „	0·063	„ 3 the vortex	0·397
„ 6 3 „ „	0·103	„ 2 „ „	0·490

Accordingly, in the four hours immediately preceding the passage of the vortex, it was 1·56 inches, and in the last two alone 0·980 inches.

dinary brightness crossed the sky, not vertically, but horizontally, flying round more than one-third part of the horizon. At every flash birds of passage on the wing redoubled their shrieks of terror. In the streets, in open rooms, and in the chimneys, the inhabitants caught ducks, quails, fieldfares, nightingales, flycatchers, and other birds.

The fall of the barometer at the different stations is given in the following table :—

Bordeaux,	from 13th to 16th,	0·743	inches, and up to the 20th rose	0·711	inches.
Toulouse	„	12th	„	0·656	„	„	„	0·640	„
Marseilles	„	10th	„	0·717	„	„	„	0·543	„
Privas	„	10th	„	0·752	„	„	„	0·555	„
Lyons	„	9th	„	0·720	„	„	„	0·524	„
Geneva	„	10th	„	0·682	„	„	„	0·492	„
St. Bernard	„	11th	„	0·715	„	„	„	0·443	„
Dijon	„	10th	„	0·693	„	„	„	0·488	„
Paris	„	13th to 15th,	0·601	„	„	„	0·507	„

According to Ehrenberg's microscopical analysis, there is nothing either in the internal or the external character of the dust which fell on that occasion which would lead him to attribute its origin to Africa. On the contrary, many of the forms therein contained were, either mainly, or in a great degree, indigenous to South America. It could not have come from the interior of a continent, but must have been carried from the sea-coast (if we may assume that it all came from one district), inasmuch as it contained existing marine forms.*

This case which we have just described forms a very natural introduction to the consideration of the storms which arise outside the zone of the Trade-winds.

---

* *Berichte der Berliner Akademie der Wissenschaften* (Reports of the Berlin Academy of Science), 1846, 227.

II.—Storms which arise at the exterior edge of the Zone of the Trade-winds.

In the preceding pages we have referred the West India hurricanes to the interference of lateral cross-currents with the upper Trade-wind on its return from the equator, portions of which, being forced to enter the lower strata of the atmosphere, meet with a constant wind, moving in a direction opposite to their own, and thus produce a cyclone. Outside the Trade-wind area the upper current descends to the surface of the earth, and is predominant there in different districts at different times, while the under-current in the opposite direction is not constant. Here, then, we shall find that the conditions of interference will constantly be presented, but the currents will be directly opposed to each other, so that they will only check each other's progress. In this case, a partial gyration indicates nothing more than that one of the currents has at last overcome the resistance of the other, a success which is just as liable to be reversed as not. In all such cases, the warm equatorial current will produce a considerable depression of the barometer, inasmuch as in its progress into colder regions it loses the moisture which it had brought with it, in repeated discharges of rain. This fall of the barometer will be observable along a line perpendicular to the direction of the current, rather than at the centre of a rotating mass of air. In the case of cyclones the height of the barometer increases in all directions from the centre; in the case now under consideration the barometer will rise to a great height over an extensive area, affected by the polar current, which has been checked in its progress. This contrast of two districts, with respectively high and low levels of the barometer, will also be marked by great differences of temperature between the regions in question.

The investigation of such phenomena requires observations from a great number of stations scattered as thickly as possible over an extensive area. In the consideration of the preceding class of storms, we have derived our information principally from the logs of ships, but in this class we are referred, by the nature of the case, principally to observations made on land. In a larger work of mine,[*] I have described at length several instances of this interference of and strife between two opposite currents, of which I shall quote a few cases here.

THE CONFLICT OF OPPOSITE CURRENTS IN JANUARY 1850.

The month of November 1849 had been very warm throughout Germany, but at the end of the month an intense degree of cold, accompanied by easterly winds, was felt in European Russia and the east of Germany. This was broken for a short time, in the middle of December, by southerly and westerly winds; but towards the end of the year it recommenced, and in January, with easterly winds, reached a degree of intensity in eastern Germany such as had not been recorded at many of the stations since the observations commenced. At Posen the thermometer fell to −37°·7 Fahr.; at Bromberg

---

[*] *Darstellung der Wärme-Erscheinungen durch fünftägige Mittel von* 1782 *bis* 1855, *mit besonderer Berücksichtigung strenger Winter* (Representation of the Phenomena of Temperature by Five-Day means, from 1782 to 1855, with a Special Reference to Severe Winters). Berlin, 1856, folio, 113 pp.

to $-33°\cdot92$ on the 22nd. The reading recorded at Vienna for the night of the 22nd and 23rd was $-13°\cdot9$, and was the lowest which had been observed since 1775.

If we examine the horizontal extension of this unusual degree of cold, we find that from its maximum in the western provinces of Prussia, in Posen, Silesia, and Bohemia, there was a continual decrease observable on all sides. In the south of Europe it was still considerable, since snow fell at Constantinople to the depth of some feet, and the temperature on January 21 was $+5°\cdot0$ in that city. At Cæsarea it was $-0°\cdot4$ on the 25th; in Simpheropol $-13°\cdot0$; and on the south coast of the Crimea $+0°\cdot50$. At Salonika men and cattle were frozen to death, and at Smyrna the winter was severer than any of the inhabitants could recollect it ever to have been. Vesuvius, and the whole range of mountains from Castellamare and Sorrento to Cape Minerva and Capri, were covered with snow, as were also Cape Bon and the Island Pantelaria; while even at Tripoli snow fell, and many of the gazelles in the desert were frozen to death.

It seems that a warmer stratum of air lay above this icy polar current, for on the day on which the minimum temperature ($-17°\cdot5$) was observed at Heiligenstadt, the temperature at the Brocken, at a height of 3,600 feet above the sea, was $+8°\cdot43$. At Schlegel in the Grafschaft of Glatz, the minimum temperature was $-28°\cdot75$ on the 22nd; at Pischkewitz it was even lower ($-35°\cdot5$); while at Wünschelburg the winter morning was so pleasant, that some people set out on a pleasure excursion to Glatz, which lies at a lower level, and could not understand how it was so cold there. The same facts were observed in Carinthia, where at Klagenfurth, at an elevation of 1,384 feet, the minimum temperature was $-18°\cdot62$;

at Sagritz, at the height of 3,590 feet, it was $+0°·5$; and on the Obir, at the level of 5,200 feet, it was $-0°·62$. Similarly, the minimum at Schöneberg, which lies at the foot of the Thurmberg, and is 785 feet above the sea, was only $-8°·5$, while at Conitz it was $-17°·95$, and at Bromberg, at a still lower level, it was $-33°·92$.

This warm air then commenced to descend; and we obtain the most accurate accounts of the prevalence of the polar current, and the conflict of the southern current with the dense cold air which lay underneath it, from the observations taken at Vienna.

On January 19, when the wind was NW., and there was a fall of snow which lay to a depth of 5·39 ft., the barometer fell rather rapidly, and the temperature was near the freezing point. At 11 P.M. on that night it began to rise from its level (28·825 inches) very suddenly, while the wind shifted to SE., and the thermometer fell to $+18°·3$ before morning. The barometer then rose until 11 A.M. on the 22nd — the day on which the maximum of the barometer and the minimum of the thermometer were so generally observed. The change of level in the two days was 1·352 inches. The wind shifted to N. and NNE.; the temperature fell to $+5°$ on the night between the 20th and 21st, and before the 22nd to $-7°·37$. Simultaneously with this change of temperature, the weather cleared up on the 22nd, and remained fine, with slight interruptions, up to the evening of the 23rd. On the 22nd, in the middle of the day, the barometer stood, for about three hours, at the height of 29·875 inches; it began, at 11 A.M. on the 23rd, to fall, at first slowly, but afterwards as quickly as it had risen before — at the rate of about 0·05 inches per hour — and continued falling up to 4 A.M. on the 24th, when it recommenced to rise. During this latter period the temperature at first continued falling, and reached the hitherto unheard-of level of $-13°·5$ during the fine night of the 22nd. The wind had remained at NNE. up to noon on the 23rd, when it shifted to N., and then to NW., bringing a really dangerous snowstorm from that point. This shift of the wind, and sudden fall of the barometer, as above noticed, was accompanied by a great rise of the thermometer, which changed, in the space of twenty-four hours, from $-10°\cdot3$ on the morning of the 23rd to $32°\cdot22$. On the 24th a decided thaw set in. The unusually heavy snowstorm on the evening and night of the 23rd was characterised by the small size of the flakes, which were more like hailstones; and it ceased with a thunderstorm at 11 P.M. In the night $7\cdot63$ ft. of snow fell, so that it attained in the city such a depth as had never been heard of.

Inasmuch as the thermometer rose $42°$ in twenty-four hours from the morning of the 23rd, it is probable that the 'small flakes like hailstones' were of a character which I have often observed, and have described ('Witterungs-verhältnisse von Berlin' [Account of the Weather at Berlin], 1842, p. 20), as follows :— ' If the southerly wind bursts in very suddenly, glazed frost (*Glatteis*) falls, i. e. rain-drops frozen while falling. In the mountains the thaw may be seen on the mountain-tops, while the cold in the valleys is still very intense. In such a case they say in the Tyrol, ' The south wind (*Föhn*) drives the cold down into the valley !'

As regards the extension of this unusual degree of cold, produced by the polar current, I must refer the reader to the 'Fünftägige Mittel,' p. viii. It will be sufficient, in this place, to exhibit the compression to which the air was subjected over the district in question. The excess of the reading of the barometer above its mean level on the 21st and 22nd, is given in the following table for the different stations, in English inches : —

Königsberg	0·884	Conitz	0·821	Breslau	0·806
Posen	0·876	Cöslin	0·819	Cracow	0·810
Stettin	0·844	Frankfort on the		Prague	0·827
Bromberg	0·833	Oder	0·804	Vienna	0·832

This shows that it was near a maximum along a line

drawn from Königsberg to Prague. From this line the excess decreased on each side to the eastward as well as to the westward, as will be seen : —

Arys	0·765	Torgau	0·729	Potsdam	0·742
Tilsit	0·710	Erfurt	0·751	Salzwedel	0·720
Memel	0·704	Arnstadt	0·654	Hinrichshagen	0·710
Warsaw	0·712	Gotha	0·629	Schwerin	0·698
Ratibor	0·762	Heiligenstadt	0·622	Lübeck	0·646
Neisse	0·633	Berlin	0·771	Copenhagen	0·702

Moving westward from the Hartz we find : —

Brocken	0·542	Canstadt	0·652	Versailles	0·566
Salzuflen	0·607	Stuttgart	0·553	Dijon	0·556
Gütersloh	0·605	Aix la Chapelle	0·643	Rodez	0·513
Paderborn	0·603	Cologne	0·605	Toulouse	0·469
Oldenburg	0·547	Cleves	0·568	Marseilles	0·611
Treves	0·661	Liege	0·633	Bordeaux	0·483
Görsdorf	0·652	Brussels	0·610	Cherbourg	0·539
Neunkirchen	0·648	Ghent	0·606	London	0·547
Darmstadt	0·648	Namur	0·551	Applegarth	0·187
Frankfort on the		Rouen	0·620	Dublin	0·519
Maine	0·666	Paris	0·566	The Orkneys	0·071

If we go southwards from the centre we find : —

Hohenelb	0·768	Pilsen	0·713	Sagritz	0·483
Senftenberg	0·706	Winterberg	0·579	Adelsberg	0·691
Olmütz	0·776	Stubenbach	0·535	Trieste	0·639
Brünn	0·760	Lemberg	0·527	Geneva	0·546
Königsgrätz	0·783	Schemnitz	0·753	St. Bernard*	0·500
Leitmeritz	0·743	Kremsmünster	0·713	Milan	0·696
Pürglitz	0·714	Salzburg	0·638	Florence	0·721
Smeczna	0·747	St. Paul	0·754	Naples	0·477
Deutschbrod	0·687	Klagenfurt	0·689		

Further to the eastward and northward we find : —

Stockholm	+ 0·281	Lugan	+ 0·279	Bogoslowsk	+ 0·151
Baltisch Port	+ 0·436	Slatoust	+ 0·224	Caesarea	nearly 0·000
St. Petersburg	+ 0·383	Catharinenburg	+ 0·157		

At Barnaoul the *depression* was 0·030 inches, at Pekin 0·167, and at Nertschinsk 0·278.

On the 22nd, the day of the barometrical maximum in Europe, the barometer in the States of New York and

* The maximum at St. Bernard did not occur till the evening of the 24th.

Rhode Island reached its lowest level; the reading at North Salem Academy on that day was −0·734.

According to the barometrical observations, of which only a graphical delineation is given by Professor Espy in the ' Fourth Meteorological Report,' there was a relative minimum on the 20th at Washita, Fort Gibson, Natchez, Memphis.

On the 21st at Fort Laramie, Fort Kearney, Beloit College, Milwaukie, Grand Rapids, Springdale, Nashville, Pensacola, Lebanon, Fort Brady, Detroit, Granville, Oberlin, New Concord, Marietta, Camden, Fort Moultrie, Chapel Hill, Toronto, Buffalo, Cattack, Gettysburgh, Seneca Falls, Harrisburg, Rochester, Newark College.

On the 22nd at Pompey, Stinnicke, Madison Barracks, Lancaster, Walterville, Burlington, Columbus, Westpoint, Albany, Newburgh, North Salem, Sing Sing, Salisbury, Ovid Plumb, Amherst, Burlington, New Haven, Sag Harbour, Wesley, Fort Adams, Southwick, Concord, Biddeford, Albion Mines.

From this we see that the minimum travelled from S. to N., whereas in Europe the maximum advanced from NE. towards SW. At Jacobshafen, in Greenland, the relative minimum occurred at an earlier period, viz. on the 21st. The district over which the rarefaction was observed did not embrace the whole continent of North America; for the reading of the barometer at Sitka, in Russian America, was 0·307 inches above the mean. The victory of the southern current caused the barometer throughout Germany to fall to such an extent, that its level had sunk 1·78 inches before the 26th and 27th throughout Prussia and Austria. The battle-field between the two currents lay now farther to the N., and the conflict is indicated by terrible snowstorms in Sweden.

Siljeström* has described this meeting of the two currents. Their alternation was marked here by similar conditions of temperature, as it had been at Vienna. At Mariestadt the thermometer rose from $-9°·4$ to $37°·4$ on the 29th, and fell again on the same day to $8°·6$. At Lilla Edet the barometer fell $1·41$ inches between the 27th and 29th, while the N. gave way to the SW. wind, the temperature at the same time rising $45°$, from $-10°·3$ to $34°·4$. When the snowstorm set in the barometer began to rise, and reached its former level on the 30th, while the thermometer sank to $1°·6$.

This was the last effort of the northern current, as it subsequently gave way to the southern current so completely, that on February 6 the barometer in Central Europe reached a level which was almost the lowest on record. In Berlin it stood, on February 6, $2·10$ inches lower than on January 21. At Stettin the difference was $2·22$ inches, and even in Flore_.__ $1·46$. The simultaneous elevation of temperature was so great, that, while the ' five-day mean ' for the period from January 31 to February 4 was $23°·9$ at St. Petersburg, $21°·3$ at Mitau, $13°·7$ at Arys, below its value, derived from the average of several years' observations, the succeeding mean, for the 5th to the 9th February, was too high to the extent of $6°·2$ at St. Petersburg, $10°·4$ at Mitau, and $9°·3$ at Arys. The increase of temperature at the three stations is found to be $25°·2$, $25°·0$, and $19°·7$ respectively. This southern current was so persistent, that at many stations in North Germany the minimum level of the barometer was not reached until the 22nd.

* *Om Snöstormen den 29 Januari*, 1850 (On the Snowstorm of January 29, 1850), in his *Afhandlingar och Smärre Uppsatser i Fysiska och Filosofiska Aemnen* (Treatises and Shorter Papers on Physical and Philosophical Subjects), 1857, p. 356.

The extension of this second minimum is given in the following table. Throughout the whole area there was a thaw, with southerly winds. The figures indicate the depression, in inches, referred to the mean for January, which nearly agrees with that for the whole year: —

Christiania	1·794	Torgau	1·189	Senftenberg	0·889
Oldenburg	1·537	Frankfort on the		Ratibor	0·947
Copenhagen	1·581	Oder	1·184	Cracow	0·903
Lübeck	1·578	Posen	1·191	St. Petersburg	0·956
Schwerin	1.534	Brussels	1·149	Versailles	0·835
Salzuflen	1·365	St. Trond	1·137	Salzburg	0·809
Gütersloh	1·340	Liege	1·134	Kremsmünster	0·812
Salzwedel	1·428	Namur	1·094	*Trieste	0·849
Stettin	1·375	Aix la Chapelle	1·084	*Milan	0·743
Hinrichshagen	1·363	Görlitz	1·073	*Florence	0·712
Cöslin	1·374	Breslau	1·077	*Naples	0·812
Cleves	1·318	Arys	1·101	Rouen	0·724
Paderborn	1·311	Baltischport	1·123	Pessen	0·768
Gotha	1·303	Treves	1·006	Schopfloch	0·730
Potsdam	1·320	Prague	1·019	Ennabeuern	0·719
Berlin	1·255	Pürglitz	0·983	*Klagenfurt	0·784
Bromberg	1·229	Smeczna	0·982	*Sagritz	0·671
Conitz	1·226	Schössl	1·043	Stubenbach	0·733
Schöneberg	1·241	Neisse	1·031	Slatoust	0·733
Dantzic	1·258	Olmütz	1·056	Cherbourg	0·626
Königsberg	1·241	Warsaw	1·036	Dijon	0·646
Memel	1·191	London	0·909	Toulouse	0·557
Stockholm	1·216	Görsdorf	0·933	Geneva	0·562
Ghent	1·190	Stuttgart	0·922	St. Bernard	0·482
Cologne	1·190	Caustadt	0·905	Bordeaux	0·316
Brocken	1·115	Pilsen	0·930	*Rodez	0·314
Heiligenstadt	1·206	Königsgrätz	0·948	Barnaoul	0·065
Mühlhausen	1·215	Vienna	0·905		
Erfurt	1·216	Brünn	0·945		

At the following stations the barometrical reading was above its mean level on the same day: —

Pekin . +0·042   Nertschinsk . +0·615   Salem (U.S.A.) +0·525

The five-day mean is relatively very high over the whole area, with the exception of the Ural, where the cold was not interrupted. At Mitau the excess was 10°·44; at Breslau, 7°·64; &c.

In contrast to the elevated temperature in Europe, we

* At the stations to which an asterisk is prefixed the minimum occurred on the 7th.

find that February 6 was the coldest day at Providence in North America, and, at the same time, that of the barometrical maximum of the year (0·829 inches above the mean height for the year). Similarly, at Cambridge, near Boston, the barometer stood 0·833 inches above its mean level, with a thermometrical minimum of −3°·1. At Savannah the excess was 0·559 inches on the 6th, and the lowest temperature 25° on the 5th. At Chapel Hill, S. Carolina, the excess was 0·533 inches, while the thermometer fell to 11°·5 on the 5th. At Muscatine in Wisconsin the barometrical maximum occurred on February 3, with a temperature of −13°·9. At Greenlake the lowest temperature was −13°·9 on the 4th. We obtain the same results from the graphical delineation in the 'Fourth Report.' The barometrical maximum appeared in the Western States earlier than in those on the coast of the Atlantic, where, however, the oscillation was more extensive. In the latter states the barometer was very low at all the stations up to the 3rd, and then rose suddenly to a maximum on the 6th. These facts give us, on a great scale, an instance of a compensation between the distribution of pressure and temperature in a horizontal direction ; and we should be able to investigate it thoroughly if the American observations had been published in a manageable form. There is more information relative to this great atmospherical disturbance to be derived from the 'American Almanack,' which contains the numerical values of the monthly means and monthly extremes of the barometer and thermometer for a few stations, than from the curves for fifty-five stations, which are printed on huge sheets, and of which there is not a single syllable of explanation contained in the 240 4to. pages of Introduction. Mr. Espy, instead of furnishing a Report, as the title-page promises, talks, as usual, of

nothing but himself and his theory. In this work, Mr.
Espy challenges me to discuss his theory, in which he
attributes the fall of the barometer to the latent heat set
free by the condensation of vapour during a *courant
ascendant* shower.   I should be prepared to do this if
Mr. Espy would first explain the fact which was established by me nine years before he propounded his
theory — viz. that the motions of the barometer and
thermometer on the W. side of the windrose are in the
opposite direction to those on the E. side — a fact which
in itself is sufficient to upset his theory ; or would even
pay any attention to my publications on the subject, which
he has never deigned to do.   The phenomena of most
common occurrence — that most extensive condensations
of aqueous vapour take place in those parts of the torrid
zone which are not visited by hurricanes, without any
change of the barometer, of any extent, being observable ;
that in the temperate zone the barometrical oscillations
are most extensive in winter, at which period the action
of the *courant ascendant* disappears entirely in proportion
to that of the ordinary atmospherical currents ; that, as
the instance just quoted proves, a barometrical minimum
may advance over an entire continent in one direction,
while a maximum is advancing over another continent in
the opposite direction ; that the yearly barometrical curve
for the interior of continents is of a totally different form
from that for the coasts of the oceans,—show us so clearly
the necessity of making a distinction between the local
courant ascendant and the great horizontal currents, whenever we attempt to discuss atmospherical phenomena, that
this must always be presupposed when we treat of any
theory which is intended to embrace the phenomena of
the atmosphere.   With reference to the fact established
by me — that the rarefaction which takes place over

Asia in the summer is the cause of the monsoon — Mr. Espy informs me, at p. 135, that the Monsoon of India has a SW. direction, and therefore does not blow towards the interior of Asia, neglecting, as he always does, to take any notice of the influence exerted by the rotation of the earth round its axis. The words of Redfield are exactly suited to this objection, who, when Mr. Hare charged me with having, in adherence to Redfield's theory, maintained that the axis of an advancing cyclone was inclined forwards, forgetting that such an inclination ' *would lift the base in the rear*,' replied, ' *It seems hardly to require an answer*.' The reason that I have never answered Hare's ' Strictures on Prof. Dove's Essay on the Law of Storms ' is, that as long ago as in the year 1828 I had published a paper on Thunderstorms, and therein given at length my reasons for considering that the electrical phenomena which accompany heavy showers were only secondary phenomena. There was, accordingly, no need for me to controvert an explanation which interprets the origin of storms, on the assumption that the atmosphere is placed between two oceans of electricity of opposite kinds, one of which is terrestrial, the other celestial.

The disturbance which we have been considering did not terminate in February, for there were violent storms in March and April. Martin has discussed them fully in ' A Memoir on the Equinoctial Storms of March—April 1850 ; an Inquiry into the Extent to which the Rotatory Theory may be applied,' 1852, 8vo.

SNOWSTORM IN DECEMBER 1850.

A good example of this type of storms is to be found in a snowstorm which has been investigated by Spassky.[*]

[*] *Note sur la Tempête d'Hiver qui a fait oeaucoup de Désastres à Kalouga, Toula, et à Kursk entre le 9—11 Dec.*, 1850 (Note on the Snowstorm which

This storm lasted for from thirty to forty-eight hours, from the 9th to the 11th of December, without intermission. A thaw had preceded the violent NW. gale; but the first gust caused the thermometer to fall about 40° Fahr. below the freezing point; so that persons caught out of doors fell dead, being buried in the drifting snow, some close to their own homes. After the storm 311 persons were found frozen to death in the government of Kaluga, 140 in that of Tula, and 39 in the district of Kursk. It is probable that many more are as yet undiscovered, owing to the depth of snow. Houses were blown down; and even horses yoked to the sleighs were frozen to death. This atmospherical revolution was the result of the struggle between the two currents of air, by means of which Professor Dove's theory enables us to explain most atmospherical phenomena; and yet many physicists continue to dispute its truth, in spite of plain proofs like the above, which cannot be explained on any other hypothesis. In order to convince us that all the characteristic phenomena of the atmosphere were in exact accordance with Professor Dove's theory during this storm, we have only to refer to the observations at Moscow. Before November 6 the polar current had prevailed there, wind N., the barometer 29·865 inches, thermometer $-0°·625$. This gave way on the 6th to the equatorial current; so that the barometer fell to 28·486 inches before the 9th, while the thermometer rose to 28°·4, and the vane shifted through S. to SW. After the polar current had yielded in this manner to the first attack, it collected its forces to repulse the foe that had forced it back. There was a lull on the morning of the 9th, to which succeeded the N. wind until the 11th, the barometer rising from 28·486 inches to 29·149, and the thermometer sinking from 28°·4 to $-6°·25$. On the evening of the 11th there was a fresh lull, on the 12th the equatorial current again, the wind shifting to SW.; the thermometer rising on the morning of the 13th above the freezing point, and the barometer sinking on the same day to 28·610 inches.

caused such Disasters at Kaluga, Tula, and Kursk between the 9th and 11th of December, 1850).

STORM IN DECEMBER 1855 AND JANUARY 1856.

In November 1855, the sirocco had been remarkably violent in the Mediterranean, and had committed great havoc, especially in Sicily, owing to the very heavy rains which accompanied it.   In a circular issued by the Governor, Prince Castelcicala, on November 22, the damage done by the inundations in the vicinity of Messina is estimated at 5,000,000 ducats, and contributions to relieve the distress are requested.   At Cantazaro, on November 17, the hurricane destroyed the mulberry plantations, as the soil was washed away from all the lands which lay close to the river.   Up to the end of November the mills were so choked with mud as to be useless, and a scarcity of bread ensued.   At Beyrie (Department des Landes), 8·5 inches of rain fell in ten days, between October 27 and November 5, being one-third of the mean annual quantity at that place.   During the storm there fell at Ancona 1·57 inches of rain on November 19, 1·85 on the 20th, and 3·11 at Bologna on the 10th. The Morea was visited by such storms that the Eurotas rose 30 feet in a night, while all the brooks became rivers and overflowed their banks.   A frightful storm raged at the Sulina mouth of the Danube from the 10th to the 13th, thirteen ships went ashore, and eight of them became total wrecks.   At Sebastopol on the 24th there was a high S. wind, the rain fell in torrents, and the roads became fathomless lakes of mud.   No trace of these violent rains is to be found in Northern Germany.   The weather had been unusually fine throughout October, and in November only half of the usual quantity of rain had fallen.   This is shown most clearly by a comparison of the quantities observed on the southern slope of the Alps with those observed in Bohemia and Gallicia.   The respective amounts, in inches, for the month of November were, at Curzola 14·37, at Ragusa 10·7, at Valona 9·95, at Santa Magdalena (near Idria) 12·96, at Trieste 7·86, at Laibach 9·51, at Santa Maria (although at such a height above the sea) 9·50, while at Prague it was 0·53, at Cracow 0·58, and at Lemberg 0·25. The first cold, between the 2nd and 6th of November, came from the W., so that snow fell in Paris before it did in Berlin, but the greatest cold came from the NE. At Smyrna, ever since the beginning of December, the rain had been falling in torrents, and the sultry weather had given rise to frequent thunderstorms, while the barometer in East Prussia rose steadily, as the wind shifted from NE. towards E., and the cold reached such an intensity that at Claussen, on the Lake of Spirding, between Lyck and Arys, the temperature from December 2 to 6 sank 31°·1 below its normal height, and the thermometer stood at −22°·2 on the 10th. On the 11th at Zechen, near Guhrau in Silesia, the temperature was −6°·4. On the 13th post carriages of all kinds crossed the Vistula and Nogat, on the ice, at Dirschau and Marienburg. The cold on the Rhine was less intense, relatively, but extended at a later period towards the W. from Pomerania, and in the middle of the month it became more moderate at all the western stations, where at the same time westerly winds prevailed. At this time the barometer at Memel rose 1·537 inches, viz. from 29·208 to 30·745, between the morning of the 16th and evening of the 19th. At Königsberg the barometer rose 1·56 inches, while the wind veered gradually from SW. towards W., NW., N., to E., and the thermometer fell so rapidly that the mean temperature of the 16th was 33°·9, of the 17th 27°·1, and of the 18th −2°·0. At Cracow the temperature fell 59°·0 (from 36°·5 to −22°·5) between the 16th and the 20th; at Oderberg 54°·2 (from 38°·7 to

—15°·5); at Brünn 51°·5 (from 41°·4 to —10°·1); at Lienz 49°·0 (from 47°·7 to —1°·3); at Klagenfurt 43°·1 (from 37°·8 to —5°·3); however, at all these southern stations the lowest temperature occurred on the 21st. This extended so far to the southward that even at Nice the temperature was below the freezing point on that day. The polar current, which had caused the barometer to rise to such an unusual height, owing to the resistance which it had met with, began now to force its way with irresistible vehemence towards the S. On the 6th the Putrid Sea, and a great portion of the Sea of Azof at Genitsche, had been frozen over, and on the 16th the Danube froze at Galatz, the temperature being —6°·2. At Odessa the thermometer fell to —26°·5; and two women, who were on their way to the town from a village in the neighbourhood, were frozen to death before they had got out of sight of their own house. On the sar e day, at Smyrna, the wind shifted from S. to N., and the thermometer fell from 65°·7 to 29°·7. From the evening of the 18th to the 21st a tremendous storm from the NE. raged in the Black Sea. Of thirty-six ships which had run out to sea from the Sulina mouth, thirteen Piedmontese, eight Greek, three Austrian, and one Tuscan were lost, with about 300 men. The fate of the other vessels was unknown at Galatz on January 7; it was only known that double the number of wrecks had occurred at other points. In the night of the 18th the thermometer fell from 47°·7 to —8°·5 in the S. of the Crimea, and stood at +5°·0 on the morning of the 19th at Sebastopol. Forty-five vessels were lost here, among which was the British steamer *Caledonia*, and the *Cortes*, an American ship. An Austrian transport, laden with cattle, ran into Sebastopol Roads, and was sunk by the fire of the Russian batteries. The crew were, however, saved and landed under the French batteries.

At Kamiesch fourteen ships went ashore, and the corpses of the crews were washed away as often as they were thrown on shore, so that it was not until the 23rd or 24th that they could be collected and buried. Even from Bombay we have accounts of a very unusual degree of cold on December 17. On the 19th there was such a storm in the British Channel that all intercourse between Portsmouth and Spithead was interrupted. The *Queen of the South*, which was bound for Malta with troops, had to come to an anchor at Spithead. An English and a Spanish steamer foundered in this storm off the coast of England. At Kilkee, on the coast of Ireland, the sea was so rough on the 20th, after the storm was over, that while a party were watching the grand spectacle of the waves breaking on the Puffing Hole Table Rocks, a sea rose suddenly and washed away two of their number, Lieut.-Colonel Pepper and Miss Smithwick, whose bodies were never recovered. In Scotland the following remarkable phenomenon was observed :—The water of the Tweed, Leithen, Teviot, Gala, and Doon fell suddenly on the 19th, and in the Doon it rose again to its former level on the 22nd, carrying with it a large quantity of ice. The 22nd was the coldest day here (temperature 22°·1). At Greenwich it was 16°·9. A similar sudden fall of the water had been observed in the Teviot in the severe winter of 1799, and again in November 1838. It had been explained by a supposed stoppage of the stream by ice. On the same day the Seine was frozen over at several points near Paris, and the reports from the south of France complain of unusual cold. We can trace, from the fact that the coldest day in East Prussia was the 20th, in Central Germany and Northern Italy the 21st, and in England the 22nd, the gradual extension of the cold towards the W. during the escape of the cold air, which had reached its maximum of accumulation on the 19th. The storm in the Channel and on the south coast of Ireland was from the SE.

In the following table I subjoin the values for the barometrical maximum and the thermometrical minimum given by the observations at the various stations through Germany and France. The height of the barometer given is the excess observed on the 19th above the mean height for the month, in the third column the lowest temperature is given, and in the fourth the date at which this

GERMANY.

	Barom. Excess on the 19th	Minimum Temperature	Date of Min. Temperature		Barom. Excess on the 19th	Minimum Temperature	Date of Min. Temperature
Memel . . .	0·852	− 4·9	18	Potsdam . . .		− 2·9	22
Tilsit	0·883	−13·5	21	Sulzwedel .	0·881	+ 0·1	21
Claussen . . .	0·711	−19·5	20	Torgau . . .	0·794	− 2·4	22
Königsberg . .	0·886	− 9·0	21	Halle	0·799	+ 0·5	22
Hela		+10·4	20	Ziegenrück . .	0·721	− 9·2	22
Dantzic . . .	0·910	+ 0·1	20	Erfurt	0·693	− 2·0	22
Schönberg . .	0·927	− 3·8	20	Mühlhausen . .	0·731	− 4·5	22
Conitz	0·854	−12·6	18	Heiligenstadt . .	0·700	− 3·1	21
Bromberg . .		− 6·3	20	Ballenstedt . .	0·787	− 6·5	22
Posen	0·846	− 4·5	22	Brocken . . .	0·548	− 3·1	19
Cöslin	0·880	+ 0·7	21	Clausthal . .	0·679	− 5·4	21
Colberg . . .	0·867	− 3·3	22	Hanover . . .	0·856	+ 0·5	21
Stettin . . .	0·963	− 2·4	21	Lüneburg . .	0·883	+ 0·1	21
Hinrichshagen .	0·911	− 6·3	21	Otterndorf . .	0·861	+ 3·0	22
Putbus . . .	0·903	− 1·8	22	Gütersloh . .	0·719	+ 3·4	21
Wustrow . . .	0·936	+ 1·0	22	Paderborn . .	0·690	+ 1·0	21
Rostock . . .	0·909	+ 1·6	22	Münster . . .	0·666	+ 2·3	21
Poel	0·910	+ 2·3	22	Lingen . . .	0·788	+ 3·2	21
Sülz	0·864	− 7·8	11	Emden . . .	0·799	+ 3·2	21
Goldberg . . .	0·918	− 2·4	11	Cleves	0·687	+ 3·2	21
Schwerin . .	0·927	+ 1·9	21	Cologne . . .	0·627	+ 3·4	21
Schönberg . .	0·936	+ 1·9	22	Crefeld . . .	0·669	+ 6·4	21
Kiel	0·929	+ 2·8	23	Boppard . . .	0·624	+ 0·3	21
Hamburg . .	0·820	+ 3·7	21	Kreuznach . .	0·589	+ 1·4	21
Breslau . . .	0·868	− 1·8	22	Neunkirchen .	0·482	+ 2·3	21
Zechen . . .	0·875	− 4·5	22	Treves . . .	0·513	+ 4·8	21
Görlitz . . .	0·774	− 0·9	21	Frankfort on } the Maine . }	0·628	+ 0·5	21
Frankfort on } the Oder . }	0·882	+ 1·9	22	Giessen . . .	0·670	− 3·6	22
Berlin	0·885	− 1·5	22				

temperature was observed.   It is only necessary to remark further, that besides the temperatures at Claussen
and Zechen, already given, the absolute minimum at
Neunkirchen was + 0°·1 on the 12th; at Frankfort on the
Maine − 1°·7 on the 11th.

AUSTRIA.

	Barom. Excess on the 19th	Minimum Temperature	Date of Min. Temperature		Barom. Excess on the 19th	Minimum Temperature	Date of Min. Temperature
Hermannstadt .	0·582	− 6·3	15	Czernowitz . .	0·933	−13·7	20
Kronstadt .	0·587	+ 1·9	20	Jaslo	0·834	−26·3	20
Zawalje . . .	0·457	+ 1·0	20	Lemberg . . .	0·866	−18·9	20
Fünfkirchen .	0·541	− 0·2	13	Rzeszow . . .	0·830	−17·5	19
Schemnitz . .	0·555	+ 1·0	19	Kesmark . . .	0·472	−22·0	20
Szegedin . . .	0·575	− 0·9	19	Cracow . . .	0·821	−22·5	20
Debreczin . .	0·519	− 3·1	19	Cilli	0·412	− 7·4	9
Klagenfurt . .	0·456	− 5·4	21	Brünn	0·685	−10·1	20
Tröpolach . .	0·466	− 7·6	21	Pürglitz . . .	0·759	− 4·5	22
Gratz	0·455	− 2·7	22	Schössl . . .	0·847	− 4·5	22
Linz	0·484	− 4·5	20	Oderberg . . .	0·769	−15·5	20
Kremsmünster .	0·602	− 5·1	21	Senftenberg . .	0·706	−15·3	4
St. Paul . . .	0·493	−17·5	19	Trautenau . .	0·734	−10·8	4
Neusohl . . .	0·768	−13·5	20	Leutschau . .	0·663	−13·0	20
Salzburg . . .	0·515	− 7·8	20	Leipa	0·696	− 7·2	4
St. Jacob . . .	0·421	+ 4·1	21	Wallendorf . .	0·610	− 4·9	18
Vienna . . .	0·692	− 2·2	20	Melk	0·627	− 2·4	20
Prague . . .	0·762	− 2·9	22	Greston . . .	0·608	− 11·2	20
Bodenbach . .	0·793	− 0·9	20	Wilten . . .	0·374	−15·7	21
Reichenau . .	0·571	−24·3	21	Kahlenberg . .	0·595	− 4·9	19
Czaslau . . .	0·713	− 4·7	20	Tyrnau . . .	0·664	− 2·7	10
Olmütz . . .	0·817	− 9·0	20	Elischau . . .	0·637	−11·4	4

BELGIUM, &c.

	Barom. Excess on the 19th	Minimum Temperature	Date of Min. Temperature		Barom. Excess on the 19th	Minimum Temperature	Date of Min. Temperature
Liege	0·595	+ 0·7	21	Stavelot . . .	0·537	− 3·1	22
Brussels . . .	0·566	+ 8·2	22	Ostend . . .	0·513	+11·1	22
Ghent	0·563	+ 5·0	22	London . . .	0·427	+16·9	22

FRANCE.

	Minimum Temperature	Date of Min. Temperature		Minimum Temperature	Date of Min. Temperature
Lille	+ 9·5	22	Vendôme.	+ 13·8	21
Hendecourt	+ 7·5	22	Nantes	+ 23·9	20
Clermont	+ 5·5	22	Grangeneuve	+ 14·0	20
Les Mesneux	+ 4·8	21	La Châtre	+ 15·8	12
Metz	+ 6·8	21	Bourg.	+ 8·6	13
Görsdorf	+ 1·4	21	Le Puy	+ 3·7	13
Paris	+ 14·5	21	St. Leonard	+ 17·6	12
Marbouó	+ 12·8	21	Bordeaux	+ 26·6	12

ITALY.

	Min. Temp.	Date of Min. Temp.
Rome .	. + 27·5 .	. 21
Bologna .	. + 7·3	
Ferrara .	. + 12·2	
Ancona .	. + 23·9	

At Lisbon the maximum height of the barometer, 0·435 inches above its mean monthly level, was not observed till the 30th, the minimum temperature 34°·9 on the 6th.   At Funchal in Madeira the maximum barometrical excess, 0·393 inches, was observed on the 23rd.

The month of December 1855, as regards the unusual height of the barometer, and the atmospherical phenomena which accompanied it, exhibits a striking similarity with the month of January 1850, which was so remarkable for its intense cold in Prussia.   The analogy was so perfect that, as in 1850, the sirocco which had lasted for a long period brought a fall of red snow in Switzerland; so in 1855 red rain, so-called 'blood rain,' was observed in the same country.   In 1850 the equatorial current, which, in its progress northwards, was forcing back the polar current, was repelled again in its turn, and did not succeed in forcing a passage until after a second slight barometrical minimum. This was also the case in 1855. The equatorial current made its appearance, on the Danube, on January 8, 1856 ; the thermometer rose suddenly from 11°·7 to 59°·0, so that the breaking up of the ice was unusually accelerated. On this day the barometer in Germany sunk to its minimum height in January, and even on the frontier of Poland the temperature rose two degrees above the freezing point. In Lisbon this minimum occurred on the 6th, and was 1·18 inches below the mean height of the preceding December. On the 6th the temperature began to fall again in Germany, and the barometer to rise, reaching a maximum on the evening of the 13th. On the same day the temperature on the Lower Danube fell from 59° to 5°, with a violent NE. wind. The change of temperature in the Crimea was just as remarkable. On the 12th, at Sebastopol, the sun was warm, and the air was delightfully mild ; on the 13th rain fell in torrents, and that evening frost set in, so that the ink froze in the pens, and the water in the barracks, the thermometer falling to 11°·1 — a change of 38°·3 in twenty-four hours. Immediately after this barometrical maximum, which was the last effort of the northern against the southern current, the latter gained a decided victory.

A spring temperature was felt over the whole of Europe to such an extent, that at Stockholm the temperature, which had been −6°·6 on the 11th, rose above the freezing point before the 15th, and in Berlin it did not once fall below it between the 16th and the 29th of the month. The point where the two currents met in the Mediterranean lay, on this occasion, more to the westward than in the preceding December ; and consequently the shipwrecks occurred chiefly on the south coasts of France and Spain. A tremendous storm raged in the Mediterra-

R

nean from the 14th to the 15th. The sea reached a great height near Cette; and the railway embankment between Cette and Frontignan was damaged to such an extent, that all traffic was interrupted. On the 16th, thirteen vessels were on shore within sight of Gibraltar. Similar accounts are reported from Cadiz, La Cortadura, Cape Trafalgar, Conil, Chichorra, and the mouth of the Guadalquivir. The English frigate Apollo was lost on the passage from Constantinople to Malta. In every place where the warm S. wind came in contact with the ground, on its descent from above, the aqueous vapour of this sirocco, which I have long ago proved to be the Trade-wind on its return from the West Indian seas, was condensed, and gave rise, as before, to great floods. At Lisbon, in Jan. 1856, 10·14 inches of rain fell in twenty-nine days. In Spain also the rain ra..ed through the whole month; so that on the 23rd, at Seville, in the Triana quarter of the town, the water reached the balconies of the houses, and the Governor went about the streets in a boat to distribute assistance. In the Tagus seventeen corpses were counted passing Aranjuez; and in Portugal the houses close to the river fell, in consequence of their foundations having been washed away. An account from Leghorn of the 12th says: 'We are here again threatened by an inundation, which bids fair to rival that of last year. The whole country from Pontendera to Empoli and Prato, and as far as Ligne, six or eight miles from Florence, is flooded. The water rose so high at Pisa as to cause very great alarm. A high W. wind prevails, the temperature is high, and there are occasional thunderstorms.'

A thunderstorm of great violence passed over Clermont, Les Mesneux, Görsdorf, and Vendome on the 24th. The Loire at Nantes was 15·55 feet above its lowest level;

and between the 24th and 29th the Garonne rose above its banks at Bordeaux, as the Lot had been swelled by constant rain. In contrast to this, only 0·51 inches of rain fell in Algiers in January, so that the dryness of that month was remarkable.

The table of the barometrical maximum on the 13th and 14th of January, at the Prussian and Austrian stations, is here subjoined. It is calculated in English inches, like that for the previous December, which has been given above. We see by it that the effect of the meeting of the two currents was more decidedly marked to the SW. than to the NE. : —

GERMANY.

Mannheim	0·632	Brocken	0·654	Putbus	0·645
Frankfort on the Maine	0·671	Clausthal	0·676	Stettin	0·679
		Ballenstedt	0·691	Colberg	0·628
Giessen	0·701	Heiligenstadt	0·630	Cöslin	0·592
Treves	0·648	Mühlhausen	0·724	Frankfort on the Oder	0·688
Neunkirchen	0·599	Erfurt	0·689		
Kreuznach	0·671	Ziegenrück	0·718	Görlitz	0·719
Boppard	0·705	Halle	0·753	Zechen	0·689
Cologne	0·732	Torgau	0·700	Breslau	0·697
Crefeld	0·736	Berlin	0·714	Ratibor	0·653
Cleves	0·771	Salzwedel	0·746	Posen	0·652
Emden	0·761	Hinrichshagen	0·645	Bromberg	0·659
Lingen	0·799	Schönberg	0·717	Conitz	0·569
Paderborn	0·732	Schwerin	0·720	Schönberg	0·643
Münster	0·691	Poel	0·689	Dantzic	0·563
Gütersloh	0·755	Rostock	0·662	Königsberg	0·492
Lüneburg	0·689	Goldberg	0·640	Claussen	0·468
Otterndorf	0·726	Sülz	0·642	Tilsit	0·484
Hanover	0·766	Wustrow	0·671	Memel	0·488

AUSTRIA.

The sequence of the stations in the following table is not strictly from south to north, as the extent of the area over which the observations were extended would not permit of such an arrangement : —

Curzola	0·589	Parma	0·663	Adelsberg	0·664
Ragusa	0·584	Sondrio	0·573	Zawalje	0·638
Zara	0·541	Milan	0·669	Fünfkirchen	0·718
Trieste	0·678	Bologna	0·634	Szegedin	0·661
Urbino	0·562	Botzen	0·639	Laibach	0·719
Venice	0·690	Meran	0·586	Wilten	0·487

Cilli	0·707	Czaslau	0·706	Czernowitz	0·585
Sta. Magdalena	0·592	Leutschau	0·626	Schössl	0·746
Debreczin	0·659	Salzburg	0·662	Obervellach	0·679
Vienna	0·704	Melk	0·672	Linz	0·555
Olmütz	0·718	Trautenau	0·625	Kremsmünster	0·682
Rzeszow	0·664	Althofen	0·547	Tröpolach	0·669
Jaslo	0·662	Lienz	0·666	Cracow	0·613
Lemberg	0·575	Schemnitz	0·601	Heiligenblut	0·460
Gratz	0·572	Leipa	0·675	Innichen	0·546
Gastein	0·321	St. Peter	0·665	Markt Aussee	0·567
Prague	0·719	Kahlenberg	0·670	Reichenau	0·688
Brünn	0·688	Pilsen	0·723	Klagenfurt	0·702
Alt-Aussee	0·487	Pürglitz	0·762	Admont	0·535
St. Jacob	0·611	Hermannsstadt	0·654	Sta. Maria	0·103
Tirnau	0·715	Neusohl	0·955		
Bodenbach	0·664	Senftenberg	0·678		

The last station, which is at a great height, seems to show that the equatorial current was never interrupted at this level, and that the polar current, blowing in the opposite direction, was only able to effe :t an entrance into the lower strata of the atmosphere. Subsequent to the maximum, southerly winds replaced those from the north over the whole district.

At the first appearance of the equatorial current, in the beginning of January, the polar current, finding its way blocked up in Europe, seems to have forced a path for itself in America. This is shown by the fact that on the 5th a storm raged, for from fifteen to eighteen hours, across the whole continent, from Virginia to California, with such violence as to stop even the railway traffic. This was followed by a very heavy fall of snow, while the thermometer fell to −15°·25.

A very characteristic and distinctive mark of the storms which arise from the interference of two opposite currents is, that the fluctuations of the barometer, which are observed when the wind changes its direction, are accompanied by great variations of temperature. On the path of an advancing cyclone the direction of the wind shifts from one point of the compass to that which is opposite to it: and after the lull, the barometer begins to rise again as quickly as it had fallen before it ; but the temperature remains unchanged, inasmuch as, in its advancing course, the air, moving in spirals, may pass, perhaps, twice over each station, over which the cyclone travels.   Consequently, in this instance, there is no connection between the temperature and the direction of the wind.   Another important difference is, that every storm in the torrid zone travels along a clearly defined track, along which no others ever return in the opposite direction ; whereas the storms of the temperate zone are often characterised by such an oscillation, backwards and forwards, as that described.   Hence it is at the very outset unjustifiable, when (in cases similar to that just considered) the currents which are striving to drive each other back produce a local whirlwind, to identify this with a cyclone.   In this case the expression, ' clear-weather side of a storm,' which is used by Redfield, may possibly indicate winds which, even if they have a rotatory motion, are yet not necessarily cyclones.   In such an instance as this the thermometer would serve as a guide, when the vane and the barometer leave us in uncertainty as to the character of the storm with which we have to deal.

THE STORM OF JANUARY 17, 1818.

I am disposed to attribute the great storm in Eastern Prussia, to which allusion has already been made (p. 157), to a strife between the polar and equatorial currents, in consequence of the following facts : —

On December 10, 1817, a very severe frost set in at Archangel, with a NE. wind, so that the thermometer fell to $-19°\cdot75$ on the 11th, and $-33°\cdot25$ on the 14th, and the barometer reached its highest level — 30·683 inches (reduced to 32°) — on the evening of the 13th, the wind still continuing in the E. and SE. points. At Dantzic the barometrical maximum — 30·214 inches — was not observed until the 17th, the wind having been at first from the N., and then from the E. The cold at Archangel was mitigated by westerly winds, which prevailed for a short period, and the temperature rose to +11°·75 by the 20th ; but as the wind returned to the NE. and SE., the temperature fell to −44°, and was −38°·9 on January 1, 1818. The barometer now began to indicate the approach of a south wind by its fall ; and this wind set in on the 3rd, raising the thermometer to the freezing point on the 4th, so that the change of temperature in three days was 70°·4 (from −38°·9 to +31°·5). This mild temperature continued, with slight intermissions, until the barometer reached its lowest level — 28·252 inches — on the evening of the 16th, 2·431 inches below the maximum of the preceding month. The shift of the wind during the hurricane of the 17th at Königsberg was S., SW., W., NW. ; and fourteen miles from that city there was a thunderstorm. The barometer fell from 30·788 inches on the 3rd to 28·879 inches on the 18th — 1·909 inches in fifteen days ; while the thermometer rose from −18°·8 on the 3rd to 34°·7 on the 4th (53°·5 in four days). The wind shifted from E. to SW. ; and the warm weather lasted so long, that we soon saw the green fields without snow. At Dantzic the barometrical maximum, on the 3rd, was 30·586 inches, and the temperature on the next morning was −4°, with a SSE. wind. In the course of the succeeding night, sheet lightning heralded the approach of the equatorial current, which came on during the night of the 6th as a S. wind, and then shifted through SSW. to W., bringing on a thaw. The barometer fell to 28·890 inches on the 17th, the wind shifting from S. to W.·; but rose to 30·152 inches by the 20th, when the wind had become WNW. Howard says that during the violent storm of the 15th, the Elbe at Hamburg rose to such a height that the lower parts of the town were flooded, and the people had to go about the streets in boats. The storm did great damage at Stettin and Königsberg. At Edinburgh there was '*a perfect hurricane*' at 5 P.M. on the 15th, the barometer having fallen 0·8 inches in the morning. At the west end of Princes' Street, the minarets of St. John's Chapel were all blown down, great and small, so that the top of the tower was a complete wreck. Howard gives the character of the month for London as '*a succession of gales.*' This wild weather did not cease at once; for Buck ('Climate and Weather of Hamburg'), who enumerates only four hurricanes at that place in the course of thirty years, mentions one on February 27, 1818, accompanied by a high tide. Howard says of February, 'Winds changeable; at the end of the month stormy from the W.' The Journal of Observations at Dantzic gives westerly winds during the whole month, with a barometrical minimum of 29·009 inches on the 26th, with a temperature of about 36° or 37°, and a high W. wind, succeeding a barometrical maximum of 30·286 inches on the 17th, showing that the struggle was not even then concluded.

It is evident that the original disturbance of the equilibrium, which we have been unable to examine more accurately, in consequence of the want of a sufficient number of stations of observation, was due to the intrusion of the equatorial current into a district in which the polar current had prevailed for a long time, and had produced a very intense degree of cold. The whole series of storms followed from this interference. We hear from Edinburgh, on January 3, 'Notwithstanding the severe and uninterrupted frost, there is very little snow lying;' and on the 14th, 'At ten o'clock on the evening of the 12th a SW. wind set in, with heavy rains, and became a perfect hurricane in the course of the night.' Farther to the south the equatorial current had set in much earlier; for while in London the temperature on the 11th was +14°·0, in the Bay of Biscay there had been a terrific storm ever since the 9th, and between the 7th and 8th twenty vessels were wrecked on the coast of France, between Brest and St. Malo. On the 15th there was a thunderstorm at Whitehaven, of a violence 'such as perhaps had never been experienced in that district.' These storms were very severe on the south coast of England. A letter from Jersey of December 19, says : 'For several days we have had continual storms, with heavy rain and hail, while the wind has been shifting about between NNW., W., and SW.'

The instance which Loomis has recently investigated ('On certain Storms in Europe and America '), of the storm which lasted from the 21st to the 28th of December, 1836, confirms, for the district of Europe, in a very intelligible manner, by means of coloured maps, the view which I have taken of such phenomena in my 'Representation of the Phenomena of Temperature by means of Five-day Means.' This storm had been immediately preceded by an analogous storm in America.

STORM IN FEBRUARY 1823.

In the case of all the storms which have been described, with the exception of that of January 17, 1818, we have found ourselves on the polar side of the line of contact of the polar and equatorial currents. If there were a series of stations extending from the northern coast of Africa into the interior, we should be in a position in which we could examine the equatorial current more closely, and perhaps follow it back to the point where it first comes down to the earth on its return, as counter-trade-wind, from the upper regions of the atmosphere. We must, however, abandon the hope of obtaining any meteorological co-operation for that region, and must therefore content ourselves with looking for an instance in which the battle-field of the currents, which is usually on the shores of the Mediterranean sea, has been, by chance, moved to the northward of that district. This was the case on the 2nd and 3rd of February, 1823, so that in this instance we are in a position to add observations of the barometrical minimum, which ensued along the northern boundary of the equatorial current, in order to supplement the maxima which have been before described, and thus to take a general view of the entire phenomenon. In the year 1828 I published my investigations upon this storm, from which it appears that the barometrical minimum had been preceded by unusually cold weather in France and Germany, which had set in on December 3, 1822. At Strasburg the frost had lasted 50 days in succession, longer than had ever been known there before, according to Herrenschneider; and at Paris there had been a NE. wind for seventeen successive days, with a high barometer. At last, on January 15, a SW. wind set in and was felt. as far as Dantzic, the temperature rising accordingly. Fresh easterly winds brought fresh cold, so that by the 24th the barometer had risen again at Dantzic, and the thermometer had fallen to $-11°\cdot9$. At Paris the barometer continued low and the sky overcast, in spite of the fresh cold, as the SW. wind still continued in the higher strata of the atmosphere. This is proved by the facts that while the temperature at Liddes in the Valais was $+0°\cdot5$ it was $+6°\cdot1$ at Martinach, which lies at a lower level; and that at Joyeuse, at the foot of the Tanargue, which stands up like a perpendicular wall,

5,000 feet high, more than an inch of rain fell from the warm upper current, at a time when the ground was frozen, and snow fell farther to the southward. On the 2nd and 3rd of February, the barometrical minimum occurred, and the depression below the mean level is given in the following table, in inches : —

Gosport	1·367	Nismes	1·376	Berlin	0·950
London	1·172	Avignon	1·376	Breslæ	1·154
Boston	0·977	Strasburg	1·332	Leobschütz	1·154
Dieppe	1·287	Kremsmünster	1·209	Altona	0·977
Zwanenburg	1·243	Vienna	1·199	Apenrade	0·817
Cologne	1·243	Prague	1·199	Dantzic	0·924
Paris	1·270	Peissenberg	1·200	Königsberg	0·808
Tübingen	1·305	St. Gall	1·181	Tilsit	0·728
Ratisbon	1·261	Milan	1·163	Mitau	0·666
Nuremberg	1·243	Mulfetta	0·977	St. Petersburg	0·373
Geneva	1·243	Jena	1·074	Archangel	0·355
St. Bernard	1·243	Ilmenau	1·110	Christiania	0·222
Joyeuse	1·287	Leipzic	1·128		
Toulouse	1·332	Halle	1·110		

At Reikiavik, in Iceland, the barometer stood 0·435 inch above its mean level.

The observations on the direction of the wind give : —

In North Germany, England, Norway and Russia, NE., and N.

In Central France and Germany a nearly dead calm; on St. Bernard, SW., on the Peissenberg, W.

At Lisbon and Constantinople fearful storms, at Genoa a very heavy sea produced by a storm.

At Joyeuse 2·4 inches of rain fell on January 11, and 1·3 on February 2. At London 0·98 inch on the 1st and 2nd.

At Avignon the Rhone rose to a level before unheard of in the month of February. This was attributed to a sudden thaw in the Alps.

The barometrical observations show two points of minimum pressure; one on the west coast of England, the other on the south coast of France, advancing from the

SW. towards the NE. The extent of the depression decreases rapidly as we go in a northerly direction, becoming insignificant in Norway and the north of Russia. After the minimum the barometer began to rise rapidly with a return of frost. At Königsberg the thermometer stood at 38°·7 on January 31, 58·5 degrees higher than it did on the 25th, when the reading was − 19°·8. On February 1, the NE. wind set in, at first gently, but afterwards grew stronger, and brought the thermometer down to − 8°·0 by the 7th, and the barometer up to 30·359 inches on the following day, 1·261 inch higher than it stood on the 3rd. At Dantzic the thermometer fell from 34°·2 on the 1st, to −5°·3 on the 8th, and the barometer rose 1·225 inch during the same period. At St. Petersburg the NE. wind continued throughout, without interruption, but the barometer rose from 29·457 inches on the 1st, to 30·407 on the 6th, and the temperature fell from +12°·7 to − 22°·7. At Archangel the barometer rose 0·721 inch, and the temperature fell from +1°·6 to − 35°·5 between the 1st and 5th.

A general glance at these observations shows us that a northerly current forced its way into the southerly one, which had been prevalent over the whole of Europe and down into the Mediterranean, and that this polar current, which advanced from NE. towards SW., by increasing the pressure along its path, had divided the barometrical minimum into two portions. Along the boundary line between the two currents we find that fogs ensued. In the Journal of the Royal Society in London we read : ' a *most intense fog on the morning of the 3rd,*' and the same was the case at Königsberg, Dantzic, Vienna, and Tübingen. The cause which disturbed the advance of the minimum, viz. the NE. wind, set in at Königsberg as early as on the morning of the 1st, but was not observed until later at the southern stations. From the barometrical observations we obtain the total pressure due to the prevalent southerly wind, and to the partial intrusion of the northerly wind, so that we find two distinct agencies simultaneously affecting the barometer; one of them advancing from SW. towards NE. and depressing its level; the other advancing in the opposite direction and causing it to rise.

The latest example of this class of storms is the violent storm of February 9, 1861, at Dublin, according to a communication which I have received from Mr. Robert H. Scott. On this day the difference of level of the barometer between Dublin and Limerick was upwards of an inch. Professor Haughton gives the following account of the storm * : —

This storm constitutes an admirable example of Dove's second kind of storm, outside the limits of the Trade-winds. In this class of storms there is a direct opposition between the SW. or equatorial current, and the NE. or polar current of air; there is generally a succession of non-cyclonic gales, NE. and SW.; and when the SW. wind gives place to the NE. there is a rising barometer and minimum temperature corresponding to the time of the storm. The following facts place the peculiar and non-cyclonic character of this storm beyond all doubt : —

In Dublin a wave of atmospheric pressure occurred, of eight days four hours duration, the two crests of the wave being —

First Crest — Feb. 1, 22 hours, 30·70 inches.
Second Crest — Feb. 10, 2 hours, 30·48 inches.

and the hollow of the wave being —

Feb. 5, 2 hours, 29·00 inches.

The gale, or storm, occurred of maximum violence —

Feb. 9, 10 hours — velocity, 24 miles per hour.

* 'Proceedings of the Royal Irish Academy,' vol. vii. part 13. I have been kindly permitted by the Academy to borrow the accompanying woodcut.—*Trans.*

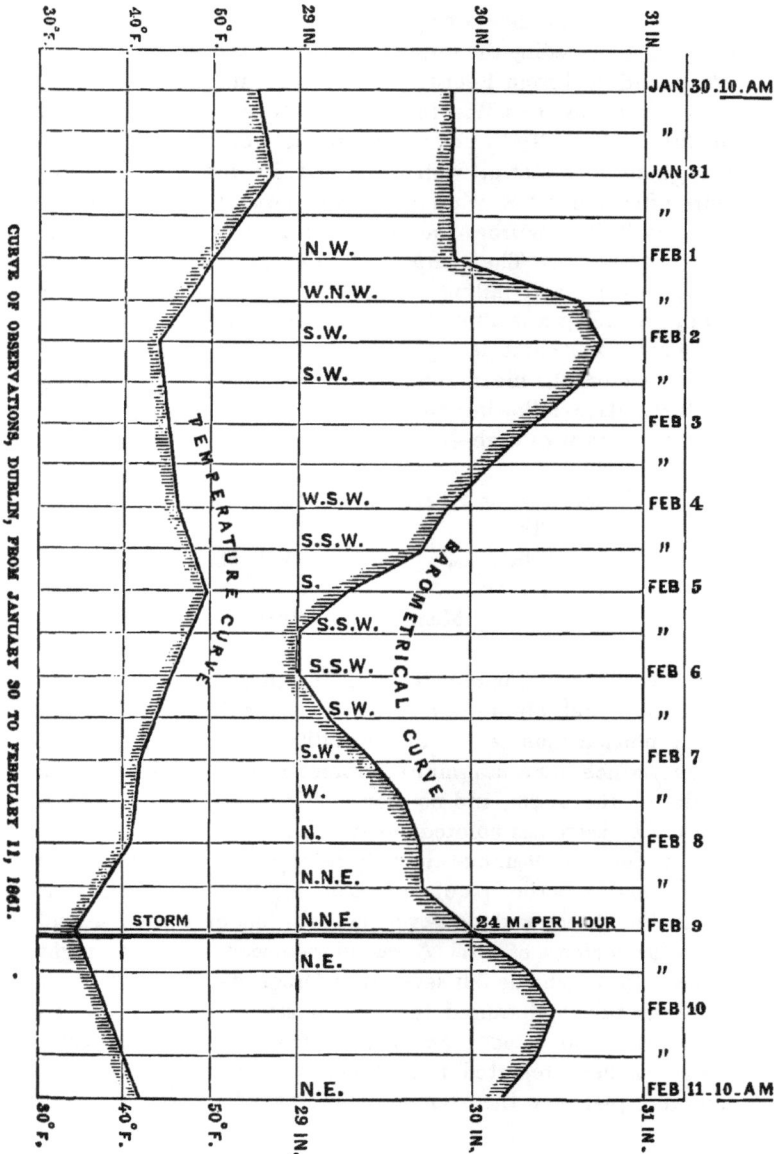

CURVE OF OBSERVATIONS, DUBLIN, FROM JANUARY 30 TO FEBRUARY 11, 1861.

The accompanying diagram (p. 253) shows all the circumstances of the week preceding the storm. We see that the equatorial wind (SW.) continued from February 2 to the evening of February 7, when it gave way to a W. wind, and finally settled, at 10 P.M. of the 8th, into a NNE. polar current, at which point it held throughout the storm, which reached its maximum twelve hours after the NNE. wind began to blow. On t.. night of February 9, it blew from the NE., and so continued for forty hours afterwards. The temperature during the prevalence of the SW. wind and falling barometer was mild, and the air damp, producing a feeling of closeness, from the vapour present in the air. For eight days previous to the minimum height of the barometer, the mean temperature was 49°·7, while, during the three days following the minimum of the barometer, the mean temperatures were—

February 7th, at 10 A.M.	. .	42°·5	
„ 8th, „	. .	40 ·8	
„ 9th, „	. .	35 ·1	Maximum of gale.
Mean	. .	39°·4	

This shows a rising barometer, a falling thermometer, reaching a maximum and minimum respectively at the time of the storm. Such a phenomenon cannot by possibility be confounded with a cyclone, which has a minimum barometer before and up to the middle of the storm, and no such relation of the gale to temperature as Dove has pointed out in the class of storms to which that of February 9 unquestionably belongs. The storm of the 9th was also only the first of a series arising from the same cause — viz. the direct and non-cyclonic collision of the equatorial and polar currents of air. A second gale occurred on the night of the 18th, which was felt severely at Drogheda, Dunmore East, and Penzance, and caused the loss of several vessels. At all three places the wind blew steadily from the SE. A third storm has been reported from London, Chichester, Plymouth and other places on the evening of February 21. It was felt in

Dublin from the SSW., but not severely.   At 7 P.M., in London, it was felt at its height; and it is said to have reached 36lbs. per foot, and it had sufficient force to blow down the spire of Chichester Cathedral.   According to Dove's theory, these two storms are supplements to the storm of the 9th, and not distinct cyclonic movements.

III. — Storms produced by the Mutual Lateral Interference of two Currents flowing in opposite directions.

If the two currents, on coming in contact with each other, have altered their paths through any angle, so that they flow in opposite directions in parallel channels, the following question arises : —What conditions will cause mutual lateral displacement after such a state of things as that described is once in existence? The most obvious cause is to be found in the fact that the cold air of the polar current exerts a greater lateral pressure than the warm air of the equatorial current, and, therefore, has a tendency to displace it. If the two currents flow in distinct channels close to each other, and if, moreover, the retardations produced by their contact have ceased, the mean velocity of both will increase. The velocity of the equatorial current will increase more rapidly than that of the polar, inasmuch as the former flows in a channel which is constantly contracting, the latter in one which is constantly expanding. Owing to the fact that the equatorial current continues for a shorter time in contact with the surface of the earth, it will be retarded by friction to a less extent than the polar current will be accelerated by the same cause ; or, in other words, the equatorial current will be more deflected towards the W., than the polar towards the E. If, e.g., the bed of the former lies in Europe and of the latter in America, the former will have a tendency to move away from the line of contact, and thereby will induce the polar current to diverge, as a NW. wind, from its own bed in order to fill the vacuum.   If the conditions be reversed, the equatorial current will have a tendency to force its way into the bed of the polar current in a more westerly direction than that of its own proper course.   This interference will take place at first in the upper strata of the atmosphere, in consequence of the greater density of the polar current.   If the resist- ance offered by the polar current be so great that the equatorial current cannot force an entry, a cyclone in the direction S., E., N., W., will be generated in the latter. In the case of a regular mutual displacement of the cur- rents, the vane will shift in the direction S., W., N., E., a gyration which is entirely distinct from that produced by a cyclone.

If we take an example, and suppose that in Europe the polar current gives way in high latitudes to the equatorial current, which is nearly due W., the former will veer gradually round to E., and we shall have in the north of Europe very mild weather, with westerly winds; while in the south it will be very cold, with easterly winds.   If the E. wind which is thus produced, be checked in its passage towards the W. by the equatorial current, which is still SW. in the Atlantic ocean, we shall have a baro- metrical maximum over the whole district where the cold E. wind is felt ; until at last it forces its way through the SW. wind over the Atlantic ocean, and the barometer falls.   Subsequently the southern current, checked in its turn, forces its way back again through the northern cur- rent, at first in the upper, afterwards in the lower strata of the atmosphere.

Under the conditions here described, the first cold will come from the NW. after a period of very mild weather, subsequently it will come from the N., and NE.   This was the case in January, 1855, as I shall proceed to show.

If a barometrical minimum in the temperate zones be compared to a long valley, a maximum to a mountain ridge, it will sometimes happen, if the area of observation be not too widely extended, that we shall find ourselves on the slope of the hill. A steeper slope than that which the observations on New Year's Day of the above named year would represent for the centre of Europe, is not often met with.

The following table gives the depression of the barometer, in inches, below its mean level for the month, on January 1 : —

	Inches			Inches			Inches	
Upsala . .	1·062		Brünn . .	0·823	2nd	Debreczin .	0·563	2nd
Stockholm .	1·613		Olmütz . .	0·836	2nd	Szegedin. .	0·590	2nd
Tilsit . . .	1·497		Wallendorf .	0·855	2nd	Ragusa . .	0·604	2nd
Königsberg .	1·436		Cilli . . .	0·878	2nd	Trieste . .	0·612	2nd
Dantzic . .	1·395		Ratibor . .	0·874		Adelsberg .	0·569	2nd
Arys . . .	1·356		Breslau . .	0·882		Meran . .	0·606	2nd
Bromberg .	1·304		Torgau . .	0·863		Laibach . .	0·615	
Cöslin . .	1·296		Lüneburg .	0·866		Sta.Magdalena	0·551	
Schöneberg .	1·239		Otterndorf .	0·865		Kremsmünster	0·601	
Conitz . .	1·214		Hanover. .	0·870		Tröpolach .	0·579	
Posen . .	1·163		Semlin . .	0·713	2nd	Sta. Maria .	0·580	
Putbus . .	1·182		Tirnau . .	0·765	2nd	Plan . . .	0·579	2nd
Stettin . .	1·130		Schemnitz .	0·750	2nd	Lienz . . .	0·562	
Moscow . .	1·121		Prague . .	0·766		Heiligenblut	0·599	2nd
St. Petersburg	1·114		Hermansstadt	0·771	2nd	St. Paul . .	0·609	2nd
Wustrow .	1·026		Czaslau . .	0·768	2nd	Klagenfurt .	0·581	2nd
Rostock . .	1·041		Bodenbach .	0·739		Obervellach.	0·589	2nd
Poel . . .	0·997		Leipa . . .	0·793	2nd	Reichenau .	0·589	
Schwerin .	0·988		Vienna . .	0·715	2nd	Trautenau .	0·537	2nd
Schönberg .	0·981		Milan. . .	0·713	2nd	Schüttenhofen	0·546	2nd
Lemberg .	1·028 2nd		Stromness .	0·755		St. Peter. .	0·609	2nd
Zechen . .	1·037		Zara . . .	0·633	2nd	Erfurth . .	0·571	
Frankfort on			Venice . .	0·662	2nd	Ziegenrück .	0·618	
the Oder .	0·993		Fünfkirchen	0·641	2nd	Gütersloh .	0·617	
Jaslo . . .	0·940	.2nd	Kahlenberg.	0·636	2nd	Paderborn .	0·605	
Czernowitz .	0·946		Linz . . .	0·627		Münster. .	0·567	
Cracow . .	0·973 2nd		Mallnitz . .	0·643	2nd	Althofen. .	0·537	2nd
Görlitz . .	0·914		Pürglitz . .	0·700		Alt-Aussee .	0·452	2nd
Berlin . .	0·893		Schössl . .	0·698		St. Jacob .	0·521	2nd
Salzwedel .	0·899		Pilsen . .	0·646		Stelvio . .	0·333	
Senftenberg.	0·896 2ud		Heiligenstadt	0·657		Giessen . .	0·501	
Kronstadt in			Mühlhausen	0·748		Frankfort on		
Transylvania	0·805 2nd		Clausthal .	0·661		the Maine	0·448	
Kesmark .	0·878 2nd		Lingen . .	0·642		Boppard. .	0·380	
Leutschau .	0·813 2nd		Emden . .	0·667		Mannheim .	0·370	

Cologne	. .	0·417	Flushing	. 0·328	Treves	. . 0·292
Crefeld	. .	0·430	Boston	. . 0·350	Luxemburg	. 0·261
Utrecht	. .	0·494	Brussels	. . 0·285	Paris	. . . 0·020
Cleves	. .	0·379	Neunkirchen	0·293	Greenwich	. 0·176
Helvoetsluys		0·410				

At Geneva there was an excess of 0·086 inches; at St. Bernard, of 0·097; at Lyons, of 0·264 ; and at Lisbon 0·388 ; the maximum during the month. In the vicinity of New York the barometer stood 0·53 inches above its mean level; at Norfolk and Charleston 0·27, while at San Francisco its level was unusually low (29·13 inches), and there was a hurricane from the SW. At Benicia there was a similar SW. storm, and the barometer stood nearly 0·62 inches below its mean level.

Throughout the whole of the Prussian and Austrian districts of observation (in the latter of which the minimum occurred, not on the 1st but on the 2nd, at all stations excepting Bodenbach, Prague, Pilsen, and Czernowitz), the year 1855 set in with a relatively high temperature, high winds from the W. and NW., and heavy falls of rain and snow. In Berlin the character of the showers was precisely that of violent thunder storms, and I expected to observe thunder and lightning every moment. They were reported to have occurred at many places between Silesia and Hamburg. Part of the island Wangeroge was washed away by the violence of the sea, and the strongest embankments along the north coast of Germany were hardly able to withstand the fury of the waves. The storm was at its greatest height at Vienna at 9 A.M. ; at Berlin at about noon. In the forest of Lambach, near Kremsmünster, where the storm was most violent at 2.15 P.M., on the 1st, 30,000 trees were blown down in a space of about 1,000 acres, 1,200 close to the observatory in a space of 300 acres. In Jaslo the roof of the court house was stripped off on the morning of the 2nd, and at Trautenau, on the southern slope of the

Riesengebirge, men and laden wagons were blown down. In Zara there was a calm.* The southern current had previously predominated over the whole area to such a degree that the barometrical mean of December had been beneath its value for seven years, from 1848 to 1854, 0·308 inches at Arys; 0·387 at Königsberg; 0·296 at Stettin; 0·309 at Berlin; 0·209 at Gütersloh; and 0·178 at Cologne. Its depression below the mean level for several years had been 0·214 inches at Cracow; 0·194 at Vienna; 0·220 at Prague; 0·236 at Kremsmünster; 0·168 at Milan. This is remarkable, inasmuch as the barometrical mean for the month of December had already been 0·178 inches too low.

A barometrical difference of 2·04 inches between Upsala and Lisbon would produce a difference of level on a water surface of 27·7 inches. It was only natural to expect, as really was the case, that the air from the area in which the pressure was not only not diminished, but even increased, would press in with great violence in order to fill the vacuum which existed over the Baltic. This cold and dense current became so universally prevalent

---

* The hourly observations give the following results at some of the stations: —

Station	Minimum Level of the Barometer	Day	Hour	Direction of Wind	Force of Wind
Prague . . .	28·544	1st	5 P.M.	W.	...
Senftenberg . .	27·626	2nd	6 A.M.	NW.	9
Cracow . . .	28·314	2nd	Noon	W.	10
Vienna . . .	28·670	2nd	5 A.M.	NW.	4
Kremsmünster .	28·138	2nd	4 A.M.	NW.	9
Salzburg . . .	28·009	1st	11 P.M.	...	...
Milan	28·969	1st	6 A.M.	N.	1

The figure 10 indicates a hurricane, the highest point of the scale of force of the wind.

that the temperature of the latter half of January and of February was unusually low.

I must refer to my 'Representation of the Phenomena of Temperature by the aid of Five-day Means,' to show the distribution of this extreme cold, and the gradual restoration of the equilibrium of the barometrical pressure.

On the very day on which this cold current forced its way into the warm atmosphere of Europe, a southerly current forced its way to the northwards in California. Between the two barometrical minima, there is interposed a district where there was a barometrical maximum, extending over the whole Atlantic ocean and embracing the eastern coasts of the United States.

A letter from Greenland says : — ' We had such fine ' warm days in February (which was desperately cold in ' Europe) and March, that we were almost tempted to ask ' if we were really in Greenland.'

An unprejudiced examination of the facts which have been stated in the preceding pages leads us to the conclusion that there are two different types of motion of the air in a storm: firstly, when its direction does not change, excepting in so far as it is modified in the course of its advance by the rotation of the earth; and secondly, when it possesses a rotatory motion about a calm centre which is advancing.   It surely cannot be denied that violent storms may be produced by the sudden irruption of the cold air of the polar current into the warm and rarefied air of an equatorial current*, or that the return trade wind, confined between meridians which are gradually

* First rise after very low
  Indicates a stronger blow.

closing in, bursts open for itself, in its fury, a channel to the north. Many of the American snowstorms appear to belong to the first of these two classes of storms, but only a very few of these have been investigated. Henry ('Agricultural Report,' 1858, p. 484) has recently given in his adhesion to this view.

In order to determine with accuracy the true character of a storm in any individual instance, it is in general requisite that we should obtain simultaneous observations from an area of considerable extent. The reason of this is that the gyration of the vane on one side of a cyclone is the same as that produced by the ordinary currents of the atmosphere, in accordance with the law of gyration. Redfield attempted to illustrate the origin of the rotatory motion of considerable masses of air, by attaching a circular disk of paper loosely, at one point, to a globe which he set in rotation. If the disk be placed near the pole it will be carried with a less velocity than if it be placed near the equator, and thus a tendency to rotatory motion will be imparted to it, which will be in the direction opposite to that of the hands of a watch, when the disk is placed anywhere on the northern hemisphere.

It is easy to see that the cause of this phenomenon is the cohesion of a solid body; a condition which cannot be applied to the particles of the atmosphere, which possess perfect freedom of motion. Taylor (Herschel's 'Meteo-'rology,' p. 67) attempts to prove that if the air rises from any point, and the air about that point flows in from all sides towards it, the result will be to produce, not as Espy and Hare assert, a direct centripetal confluence, but a whirlwind in the sense of the cyclone theory. Ferrel also agrees with this view (p. 43).

We may admit that the conditions which predispose the air to rotatory motion may be very various, but the fact which has been already proved, viz. that the gyration of the wind with the sun is the most prevalent form of its motion on both hemispheres, shows that cyclonical movements are not the predominant form of atmospherical motion. We must further remember that in the case of cyclones, we have to explain, not only their rotation, but also the fact that their centres advance in a definite direction, and turn almost at right angles to their former path, when they reach the edge of the torrid zone. There is still another point which requires explanation, which is that, although the conditions for a local *courant ascendant* are presented in abundance in all hot countries, we find that cyclones are especially prevalent in certain parts of the globe, and in these are most common at certain seasons. The very nature of a *courant ascendant* presupposes the existence of some cause which imparts a tendency to the air to rise at one point more than at any other in the neighbourhood, whether the phenomenon be due to heat or to the condensation of aqueous vapour. It seems, however, unjustifiable to suppose that such a predisposing agency could travel for days and nights along a definite path, so that it is hardly natural to post a *courant ascendant* in front of a cyclone in order to lead the way for it. The idea of the advance being due to impulsion is as unnatural as that of its being due to attraction, for the path of the cyclone crosses that of the trade winds at an angle which is nearly a right angle. Thom says expressly that the violence of Typhoons is mitigated during the day time, at which time the conditions for a *courant ascendant* are presented on a greater scale than during the night, at least as far as all the experience we have on the subject would teach us.

In addition to this, the difference in the velocity of rotation between different portions of the air of the torrid zone, which have been previously in contact with the surface of the earth, if they have not come from latitudes which are very wide asunder, is much less than that which will be observable when the air of the counter current, which has travelled from distant regions without touching the ground, comes in contact with the current beneath it. Lastly, it is not possible to explain, inasmuch as the conditions for a *courant ascendant* exist preeminently in the torrid zone, if the cyclones are accordingly peculiar to that area, why we do not ever find them within a certain definite distance of the equator. According to my theory, this is explained by the fact that there is no reason why the air should descend within the Zone of Calms, where it is ascending *en masse*, and that such a reason does not present itself until we reach a spot where the direct and counter trade winds are flowing one above the other, and interfering with each other, whenever the counter current is checked in its path, either by a lateral current or by any other cause.

I am not, however, prepared to maintain that all West India hurricanes owe their origin to the intrusion of the upper counter trade wind into the lower current, for the mechanical proof of the rotation which has been given (p. 182) holds as true as it did before, if we assume that a portion of the SE. trade wind travels far over the equator into the northern hemisphere, and gives rise to the cyclone.

The observations from New Mexico and California, which have recently been published, exhibit so complete an absence of rain in summer, that circumstances of relative dryness similar to those in the deserts in the interior of the continents of the old world, seem to exist. Barometrical observations at San Sacramento, carried on for three years, have given the following results. The table

is similar to those at pp. 52 sqq., and it exhibits a
periodical rarefaction in the summer months.

	Barometer	Dry Air	Tension of Vapour
January	+ 0·145	+ 0·242	0·262
February	+ 0·071	+ 0·172	0·258
March	+ 0·061	+ 0·104	0·316
April	− 0·025	+ 0·013	0·321
May	− 0·052	− 0·056	0·363
June	− 0·162	− 0·306	0·503
July	− 0·039	− 0·149	0·469
August	− 0·035	− 0·132	0·451
September . . .	− 0·092	− 0·143	0·410
October	− 0·055	− 0·063	0·367
November	+ 0·045	+ 0·057	0·347
December	+ 0·135	+ 0·255	0 230
Year	30·012	29·653	

It is possible that this district may exert an attractive action on the SE. trade wind, analogous to
that which, on a much greater scale, gives rise to the
SW. Monsoon in Asia. There is this difference between
the two cases, that in Asia the SE. trade wind encounters
along the whole of its front an atmosphere which has
been subjected to rarefaction, while in America such an
atmosphere is presented to the trade wind along its
western flank only, so that the hurricanes would be especially prevalent on this side. The intrusion of the SE.
trade wind into the area of the NE. trade would be, according to this view, the cause of these hurricanes.
Whether we explain them on the hypothesis of a lateral
current, which rises over Africa, and checks the counter
trade wind as it crosses its path, or on that of an advance
of the SE. trade wind along the western shore of the
Atlantic ocean, we shall arrive at the same conclusion, viz.
that they must be most abundant in the summer months.

In all storms, whatever be the form which they present,
the first problem which we have to solve is to ascertain

the cause which has occasioned the disturbance of
equilibrium; the second, to ascertain how the equilibrium
comes to be restored. In this case, phenomena may
occur which bear some resemblance to an undulatory
motion. It is, however, very unnatural to suppose that
every disturbance of the atmosphere which gives rise to a
barometrical maximum or minimum advancing over the
earth, is an atmospherical wave, as is clearly proved by
the storm of February 2 and 3, 1823 (p. 248). For the
same reason, a so-called atmospherical wave, i. e. a change
of pressure marked on the barometer by its rising from a
minimum to a maximum and falling again to a second
minimum, as the result of the displacement of an equatorial by a polar current, which again is displaced in its
turn by the former, cannot be considered as a complete
phenomenon. It will be sufficient to quote the words of
Radcliff Birt, in order to show that this conclusion has
not been arrived at solely from my investigations of great
barometrical oscillations, but is the natural result of an
unprejudiced examination of all such phenomena. In
his 'Report on Atmospheric Waves,' 1846, Birt says,
'Upon a very careful perusal of the paper by Mr. Brown,
I found that Professor Dove's theory of parallel currents,
or alternately disposed beds of oppositely directed winds,
throw a clear light on the real character of atmospheric
undulations.' Five years after the publication of this
work, and twenty-three years after the appearance of my
earlier papers, Quetelet arrived at the following conclusions:—'*Les directions des vents n'ont pas de rapports
apparents avec les directions des ondes barométriques*'
(the directions of the winds have no apparent connection
with the direction of barometrical waves). I can fully
understand why this self-evident negative conclusion is
denoted as an 'important fact.' There is, as it appears

to me, one question which those physicists who consider
the oscillations of the barometer as primary waves have
not yet answered, viz. why so great waves as these are
never traced into the torrid zone, and why the oscillations which appear at certain points within that zone, in
the region of the hurricanes, assume a totally different
form. It is to be hoped that M. Lamont will explain this
fact, as he has published in his ' Results of Meteorological
Investigations,' in the year 1857, his views on this subject.
He says, ' Inasmuch as great waves, analogous to sea-
waves, are propagated through the atmosphere, its
pressure is constantly liable to alteration, and it increases
as a wave approaches. It is generally considered that
the fluctuations of the barometer are due to variations in
the density of the air, and that the upper surface of the
atmosphere does not either rise above or fall below its
original level. I am disposed to consider that in the
course of long continuance of a current in the atmosphere its surface becomes uneven, being raised in parts
and depressed in others.' Herschel, who first suggested
the means by which such investigations have been rendered possible, viz. the simultaneous observations on the
so-called ' *term days*,' has drawn (Meteorology, p. 70) the
following distinction between the atmospherical waves
and sea-waves : — ' That the disturbances in which
they originate are not (as in the case of the sea-waves)
merely superficial, but extend through the whole depth
of the atmosphere, and are most powerful at the ground
level.' Lamont, on the contrary, transfers the origin of
barometrical oscillations to the upper surface of the
atmosphere, a region utterly inaccessible to observation.
Lastly, Admiral Fitzroy, in the appendix to the English
translation of the first edition of this work, speaks of
waves in general at p. 87, after he has discussed the

great waves, in the following words :—'All cases of apparent atmospheric waves on a smaller scale may be similarly explained by application of "Dove's Law of Gyration."'

The following statements will be sufficient to explain in a few words my views on this much-vexed question. It is erroneous to suppose that, in the case of extensive barometrical oscillations, the air *always* flows towards the point where the pressure is least ; or, in other words, that the wind is always due to a disturbance of barometrical equilibrium.   Such an interpretation would hold good neither for the cyclones, the direction of whose advance is at right angles to that of the prevalent wind, nor for the storms of the temperate zone, in which, when the equatorial current is forcing its way onwards, the barometrical minimum is advancing in the direction of the current, and the air, accordingly, is flowing towards a district in which the barometrical level is relatively higher than in that from which it has travelled.   The falling side of a great barometrical wave represents the action of this equatorial current, which may itself exert an action in bringing on the polar current to succeed it by the rarefaction which it has produced, and the diminution in the tension of the vapour which it contains, owing to the partial removal of that substance by condensation.   We see, therefore, that the rising side of a barometrical wave owes its origin to causes which are diametrically opposite to those which produce the falling side of the same wave. In this respect the great barometrical oscillations of the temperate zone differ from the local oscillations which are peculiar to tropical regions.   In the latter, when a cyclone passes over the point of observation, the fall of the barometer as the minimum approaches, and its rise when the centre has passed by, indicate the two sides of one

and the self-same storm : in the former, which are due to
gales, not to cyclones, the falling side of the wave belongs
to a different current from that which produces the rising
side.   Accordingly, the oscillations of the temperate zone
are accompanied by great fluctuations of temperature, a
characteristic which does not appear in the torrid zone to
the same extent.   The return to the ordinary conditions
of the atmosphere, as soon as the disturbing agencies
have ceased to act, takes place according to the laws
by which equilibrium in an elastic fluid seeks to restore
itself when it has been disturbed.   If these are to be
characterised as waves, it must be clearly understood
that they are not a primary phenomenon, but only one
which is secondary.

In treating of the storms of the temperate zone, we
have frequently observed that storms which have travelled
over Europe have been succeeded by storms in America ;
or that the latter have preceded the former.   Although
we are hardly justified in supposing that there exists
between these phenomena, which are observed in localities
so widely distant from each other, a direct relation to an
extent which would allow us to consider them as a
connected whole, it is evident that a disturbance of
equilibrium, which takes place at any one point, may
indirectly give rise to other disturbances which will ultimately affect distant regions.   In North America, the
mean direction of the wind in winter is NW., in summer
SW. ; the converse is the case in Europe.   This shows us
that it is not improbable that the storms of the United
States which blow from the foot of the Rocky Mountains
between November and March, and are preceded by a
rise of the barometer, equilibrate the warm currents
which advance from SW. towards NE., accompanied by
a rapid fall of the barometer, which traverse Europe at

the same period, and produce the predominant southerly direction of the wind generally observed in this continent at the end of autumn. Phenomena which have a tendency to compensate for each other, like those which have just been described, must necessarily arise on the supposition that the whole descending counter trade wind does not enter the temperate zone simultaneously along the entire extent of the edge of the trade wind zone; but that while this is the case at certain points, at others there must be an influx from the temperate into the torrid zone. Any increase in the intensity of efflux must necessarily have more or less of a tendency to generate a corresponding increase in the intensity of influx, in order to compensate for it.

I am not unaware that many natural philosophers are of opinion that the currents in the temperate zone do not correspond with the description that I have given of them, and that they imagine that the atmosphere of the trade-wind zone moves in a vertical circle by itself, and that of the temperate zone in an independent circle rotating in the opposite direction. According to this view, the district of increased barometrical pressure, at the edge of the trade-wind zone, represents the district where both the upper currents, that of the torrid and that of the temperate zone, descend to the surface of the earth. In answer to this we may remark, that the constancy of chemical composition which is exhibited by the atmosphere, is most naturally explained on the hypothesis that the energetic vegetation of the torrid zone liberates the oxygen which is requisite for the conservation of the animal life in extra-tropical regions; so that the torrid zone furnishes a momentum which is absolutely necessary for the working of the great vivarium existing on the surface of the globe. In addition to this, we find that the amount of rain which falls at the outer edge of the torrid zone could not be furnished by a current which had travelled in the upper strata of the atmosphere from the pole to the tropics.

An attempt to reconcile both these views may be made by assuming that the upper counter trade of the torrid zone descends at the outer edge of the area; flows into the temperate zone; rises again when it comes into high latitudes; flows back as a polar current in the upper strata of the atmosphere of the temperate zone; descends afresh at the tropics; flows in towards the equator, along the surface of the ground, as the ordinary direct trade wind, and at the equator rises again. According to this view, the paths of the currents may be represented by a horizontal *eight* (∞), and they would cross each other at the district where the barometer is highest, near the tropics. Maury has repeatedly given a graphical representation of this view. We know that oscillatory motions can be propagated across each other; but this could not take place in the case of currents, as whenever two currents meet, at an angle more or less acute, a resultant motion must arise, which will be intermediate between them. Maury is led by this interpretation of the question to suppose that the air at the poles moves in a vertical *eight* (8), an assumption which appears to be entirely arbitrary. In addition to this, such a view will hardly explain the fact which I have pointed out, viz. that the non-periodical fluctuations of temperature are always compensated for, within their own hemisphere, by the simultaneous existence in adjacent localities of an unusually low and an unusually high temperature of the atmosphere; so that the total amount of heat is constant, and it is only its distribution which varies at different times. This fact is easily explained

according to the views which I have proposed.   The
counter trade wind, on reaching the surface of the earth,
regains its original temperature in consequence of the
reduction in volume which it undergoes in its descent;
and, having been restored to its original state of saturation by the abstraction of the aqueous vapour which has
been condensed during its ascent between the tropics,
flows at certain points into the temperate zone.   Simultaneously with this, at other points, there is an influx of
air from the temperate into the torrid zone.   As regards
the rainfall, there exists a similar compensation on a great
scale, as may be seen from the following facts:--The
years 1857–1858 were so remarkably dry throughout
central Europe, that the Seine in Paris was lower than it
had ever been before, and portions of the bed of the
Rhine, which had never been left dry before, were un-
covered.   During the same period, the level of Lake
Ontario was two feet higher than its mean level, determined from 14 years' observations.

When we were discussing the polar current of the
temperate zone, we were led to the conclusion that it
extended throughout the entire depth of the atmosphere,
when it had succeeded in displacing the equatorial current.
When, under such circumstances, it enters the torrid zone,
it is evident that the counter trade wind will have to
make way for the upper portion of the polar current, and
to find a passage for itself into the temperate zone at
some other point.   In this case there would be a northerly
wind in the upper strata of the NE. trade wind, and this
Lartigue says that he has himself noticed.   He says,
p. 18, 'When polar currents, from N. and NE., join on to
the trade wind, the upper portion of the current has a
more northerly direction than that which is beneath it,
and the clouds, if there be any, will drift in a direction

between N. and NE.   Whenever such a union as this does not take place, above the NE. wind we have an E. or SE. wind, above that again a SE. or S., and lastly, in the highest regions of the atmosphere, a S. or SW. wind.'

General Conclusions.

IF we collect the facts which have been recorded in the preceding pages, we shall find that they will furnish us with certain general laws.

1. All permanent wind currents are deflected, owing to the motion of the earth on its axis. The equatorial currents are deflected to the westward ; the polar to the eastward. The NE. and SE. Trade-winds are permanent polar currents. The Monsoons are polar and equatorial currents alternately, according to the season of the year ; so that their directions on the northern hemisphere are NE. and SW., and on the southern SE. and NW.

2. If masses of air, originally at rest, set themselves in motion along the meridians, they will cause the vane to shift in the following manner, viz. : —

Polar currents in the northern hemisphere from N. towards E.

		southern	„	S.	„	E.
Equatorial	„	northern	„	S.	„	W.
„	„	southern	„	N.	„	W.

Hence, in general, on the northern hemisphere : —

The winds between N. & E. are the polar current

„ E. & S. are its transition into the equatorial

„ S. & W. are the equatorial current

„ W. & N. are its transition into the polar.

On the southern hemisphere:—

The winds between S. & E. are the polar current

       „   E. & N. are its transition into the equatorial

       „   N. & W. are the equatorial current

       „   W. & S. are its transition into the polar.

From this the following rotations are deduced : —

For the northern hemisphere $\rightarrow$ S., W., N., E., S. $\rightarrow$ with the sun.

  „   southern   „ $\rightarrow$ S., E., N., W., S. $\rightarrow$ with the sun.

3. A permanent wind current may be prevented from assuming the divergence which would be imparted to it by the rotation of the earth :

α. By a wind which blows constantly in a direction at right angles to its own primitive direction. These are the West Indian hurricanes, which accordingly, in the earlier part of their course, travel from SE. towards NW., and the hurricanes of the southern hemisphere, which travel from NE. towards SW.

β. By a current which has undergone less divergence. This is the origin of the Typhoons during the SW. Monsoon, a current which is bounded on its eastern side by the S. Monsoon. It is, however, probable that many of the Typhoons which travel from W. to E., owe their origin to the fact that the heavy air of the Trade-wind region, which lies farther to the eastward, forces its way laterally into the rarefied air of the SW. Monsoon, and thus generates a cyclone.

γ. By a mechanical obstruction, e. g. the storm described by Piddington and referred to at p. 189, *vide* Chart V.

In these cases, when the storm which is generated belongs to the equatorial current, the cyclones in the northern hemisphere possess a rotation which is retrograde, i. e. opposite to that of the hands of a watch; in the southern, one which is direct: hence we derive the following rules for the revolution of the vane:—

In the northern hemisphere:—

a. If the centre of the cyclone passes to the westward of a given station, the rotation will be

→ S. W. N. E. S. → or, with the sun.

b. If the centre passes to the eastward of a station, the rotation will be

→ S. E. N. W. S. → or, against the sun.

Conversely, in the southern hemisphere:—

a. If the centre passes to the westward of a station, the rotation will be

→ S. E. N. W. S. → or, with the sun.

b. If the centre passes to the eastward of a station, the rotation will be

→ S. W. N. E. S. → or, against the sun.

We see, accordingly, that in both hemispheres an equatorial cyclone passing to the westward of any station produces a rotation of the vane which is normal, i. e. in accordance with the Law of Gyration; whereas one passing to the eastward of the same station produces a rotation of the vane which is abnormal, i. e. contrary to the above-named law. Polar cyclones would produce actions the direct converse of those which have been described. If any such cyclones exist, the rotation is normal when they pass to the eastward of any station, abnormal when they pass to the westward of it.

The old rule, that complete abnormal revolutions take place in stormy weather, is thus proved to be correct. If we only know the direction in which the shift takes place we cannot decide whether there is an equatorial cyclone passing on one side of us, or a polar on the other. This would be indicated by the point of the compass at which the shifting commenced; as in the case of a cyclone, this can never be at a greater distance than half a complete revolution from the point where it terminates.

4. The vane may veer from one point of the compass directly to the opposite point:

α. When two ordinary winds, blowing in opposite directions, meet; or, as sailors say, fight with each other.

β. When the centre of a cyclone passes over the point of observation.

5. There are cases possible in which a storm, though really a cyclone, will not cause the vane to shift. This happens when the station is situated on the curve formed by the successive tangents to the path of the cyclone. In such a case the storm appears on one side of its path to have had a retrogressive, on the other a progressive, motion. If, for instance, a cyclone which possesses a retrograde rotation advances from SW. towards NE., we shall find that for the district lying on the north-west side of the path the NE. wind will be first felt at the southern stations; while for the district lying on the opposite side of the path the SW. wind appears actually to advance in the direction in which it blows, and is not felt at the northern stations until after it has been felt in the south. The first case is the well known one noticed by Franklin on the occasion of an eclipse of the moon (p. 204). Those persons who, resting on this and analogous instances, divide the wind into *positive* and *negative*—winds *par aspiration* (produced by attraction) and *par impulsion* (by impulsion)—term one side of a storm positive, and the other side of the same storm negative.

6. In the temperate zone some storms, in addition to those which have been described, arise from the lateral interference of two opposite currents. In such a case the

rotation of the vane, during the process of the interference, may be either with or against the sun, according to the directions of the currents which are displacing each other.

7. Local phenomena, such as land and sea breezes, eddies in valleys, littoral deflections of the Trade-winds, Trombs, &c., affect the vane according to circumstances which are strictly local, and are as likely to give rise to rotations in accordance with the law as against it.* In open countries they make their appearance, in the diurnal period, in a more marked way, in proportion as the regular wind-currents are less prevalent. They are, accordingly, observed in the region of the 'variables' between the Trade-winds; at the ' change of the Monsoon' in the region of the Monsoons; and, in general, in summer, when the *courant ascendant* weakens the influence of horizontal currents. It is not absolutely impossible that, independent of local circumstances, the daily apparent motion of the sun may produce a periodical rotation of the vane. If the point where the temperature reaches a maximum in the daytime be a centre towards which neighbouring masses of air are attracted, it is evident that in the morning hours the wind will be more westerly, and in the evening hours more easterly; so that the vane will move in a direction opposite to the motion of the magnetic needle.

* These small whirlwinds (Trombs) do not exhibit any regularity in the direction in which they revolve, so that several may be seen at the same time rotating in different directions. Bridgeman says that at Goruckpore, 'they turned indiscriminately, some one way and some the other. They depend on no rule producing uniformity of motion.' Reid says the same of the Bermudas, and Thom of the dust whirlwinds in Scinde. The latter says expressly that he saw half a dozen at once, and that two of these, which were close to each other, revolved in different directions. The nature of the substances which are carried up by such Trombs depends on the character of the surface of the earth: it is water in waterspouts, sand in the deserts, dust in a dry country where there is no sand, and, lastly, ashes in volcanic districts, as in those which have been described as occurring in Iceland.

If cyclones be not confined to certain districts, the chances that any given station will lie on the east side of their path are just as great as that it will lie on the west side of it. Even when the cyclones owe their origin to fixed local conditions, and exhibit consequently great uniformity in the direction of their path, it is impossible that any rotation of the vane, either in one direction or in the other, can be considered as the predominant one for stations chosen at random over the surface of both hemispheres; inasmuch as there is as great probability that a given station will lie on one side of their usual path as on the other. The prevalence of a rotation of the vane in one direction (with the sun) is therefore a phenomenon which has no connection with the rotatory motion of tempestuous winds, but is solely due to the influence of the earth's rotation on the ordinary atmospherical currents.

As regards the reason of this prevalence of one direction, there are three causes which may be assigned for it.

1. Either that all rotations of the vane through large arcs are due to cyclones. In this case —

α. If these cyclones arise in different localities, without any regularity in their appearance being observable, there will be no prevalent direction of rotation of the vane.

β. If they are local as to their origin, and traverse paths which are more or less constant, there will be certain stations on each hemisphere where the rotation S., E., N., W., is the prevalent one, and others where the converse rotation is most common.

2. Or, all rotations of the vane arise from the alternations and mutual displacements of the ordinary meridional currents, according to the principle first laid down by Hadley for the Trade-winds. In this case the rotation on the northern hemisphere is S., W., N., E., S., on the southern S., E., N., W., S.; i. e. in both cases 'with the sun.' Under

these conditions a shift in the opposite direction can never exceed a quadrant of a circle.

3. Or, lastly, rotations arise from both the causes above enumerated, viz. alternation of currents and cyclonical movements. Hence we see that rotations in both directions must exist on both hemispheres, but that those ' with the sun' must be most frequent. This is evident, because the first cause produces no rotations but those which are ' with the sun,' while the second produces as many 'with the sun' as 'against it.' Accordingly, no matter how abundant cyclones may be, there must always be more direct rotations than retrograde ones.

The previous investigations have proved that cyclones are especially prevalent in certain regions; that they follow certain definite paths, and are known under the names of —

a. The West India Hurricanes (*Aracan* or *Huiranvucan* of the coast of Mexico, *Vuthan* in Patagonia), which appear at the inner boundary of the NE. Trade-wind, and also within the district affected by that current, especially at the end of summer and in autumn, according as the original cause of the rotatory motion is due to the interference of the SE. Trade-wind with the NE. Trade, or to the intrusion of the counter Trade-wind into the true Trade-wind which lies underneath it. They travel from SE. towards NW. as long as they continue in the torrid zone, and, on crossing the boundary of the Trade-wind, turn at a right angle and advance from SW. towards NE., while the cyclone, which is rotating in the retrograde direction, expands enormously.* Cyclones appear to be less common in the SE. Trade-wind.

---

* According to Redfield, the diameter of the cyclones in the vicinity of the West Indian Islands varies from 100 to 150 nautical miles, and their subsequent expansion takes place on such a scale that the diameter often

b. The Typhoons in the north of the Indian Ocean and in the China Seas. They are most frequent in autumn, but are also very violent at the beginning of the SW. Monsoon. Their path is more from E. towards W. than from NE. towards SW., especially on the south coast of China. In the southern portion of the Indian Ocean they are also very violent, travelling from NE. towards SW., and turning at a right angle, on leaving the Trade-wind zone, so as to advance from NW. towards SE. The rotation of these cyclones is opposite to that of the hands of a watch on the northern hemisphere, and in accordance with that motion on the southern. The causes which give rise to these storms are the contiguity of the districts of the Trade-winds and the Monsoons, and the displacement of one Monsoon by the other.

The fact that cyclones are phenomena whose origin is strictly local renders it improbable that the rotatory motion which they possess is the sole, or even the most usual, type of atmospherical motion, and that it is consequently due to some cause, which is either of unknown origin, or else, as is frequently supposed, is electro-magnetical in its nature, which generates in cyclones a tendency to a rotation from right to left on the northern hemisphere, and from left to right on the southern. This is a very general impression, and has arisen from viewing the shiftings of the vane, which were produced in accordance with the Law of Gyration, as indications of a cyclonical disturbance; so that this type of atmospherical motion

reaches the extent of 600 to 1,000 miles. Thom gives their diameter in the southern portion of the Indian Ocean at from 400 to 600 miles. Piddington gives it at 240 miles in the Arabian Sea, and at 300 to 350 in the Bay of Bengal. In this latter district the diameter of the cyclone is sometimes reduced from 300 or 350 miles to 150, its violence increasing proportionately. The same author gives the diameter of the Typhoons in the China Seas at from 60 to 80 miles on an average.

has been considered to be of much more common occurrence than experience would justify us in admitting to be the case.

The correspondence which exists between the phenomena of cyclones and those produced in accordance with the Law of Gyration by the ordinary wind currents is so great, that when I published, in the year 1827, the observations which I had made at Königsberg, I drew especial attention to the following fact : — Whenever the equatorial current sets in suddenly and is as suddenly displaced again, at the precise time when the barometer is at its lowest level, the showers which belong to the equatorial current are separated from those produced by the intrusion of the polar current into it by a short period of clear weather, to which I gave the name of 'clear interval' (*trennende Ilelle*). At the centre of a cyclone, when the barometer is lowest, a similar fine moment is so frequently observed that sailors have given it a special name, 'the eye of the storm,'—*Abra-ojo* in Spanish, *abrolho* in Portuguese. Captain Salis describes its occurrence in a cyclone experienced by the ship *Paquebot des Mers du Sud*, in latitude 38° S., longitude 22° E., in these words : — 'While there was a dense bank of clouds all round us, the sky over our heads was quite clear, so that we could see the stars, one of which, right over the head of the foremast, was so bright that everyone on board noticed it. The barometer was 27·79 inches English.'

On the other hand, it must be admitted that under certain local circumstances a cyclone may be modified to such an extent that in particular localities it assumes the form of an ordinary gale ; but if its course over a large area be examined, it exhibits itself as a true cyclone. This is the case, according to Redfield, with the *Nortes* of the

Gulf of Mexico, which blow at Vera Cruz, especially from September to March. They reach their greatest height in four hours from their commencement, and then last for forty-eight hours with undiminished violence. This long continuance of the storm in the same point is explained by Redfield, on the hypothesis that cyclones, in their advance from the eastward, encounter the high land in front, so that their circular form is converted into an elliptical one, which becomes ultimately nearly rectilinear. Vice-Admiral Adam gives the three following signs of a *Norte* for the Bar of Tampico : —

1. The atmosphere is generally damp.

2. The Peak of Orizaba is clearly visible, while the lower part of the mountain is enveloped in clouds.

3. The mountains which lie inland, towards the southeast, are unusually clear, and all the animals appear to succumb to the excessive heat.

The NE. and NNE. storms of the Pacific Ocean, which occur at Guatemala and Nicaragua, and are known under the names of *Papagallo* and *Tehuantepec*, are, according to Redfield, the fair-weather side of cyclones; while the SW. gales in August and September, which are called *Tapayaguas*, represent the opposite side of similar storms.

In order to see our way through this medley of phenomena, which are identical in the forms which they present, however they may differ as to their origin, we have discussed separately, in the first instance, those which depend on the Law of Gyration, in the second, those which are due to storms, whether the storms be hurricanes or gales. I shall, in conclusion, sum up the results of the indications of the meteorological instruments in the form of practical rules. These rules will be simplest in the Tradewind zone, because that in this region the distance between the commencement and termination of the current remains constant, so that the rotation of the earth does not cause the vane to shift regularly, but only to assume a constant deflection, indicating a permanent current — the Trade-wind ; and also, the character of the storms is exclusively rotatory. In the Monsoon district they present a little more complexity, because that, in this region, the annual alternation of two currents produces an annual rotation of the vane, indicating periodical winds; and also the storms, though they are still rotatory, do not travel along so uniform a path, though this must still vary to a certain extent from the direction of the prevalent wind in order to produce a cyclone at all. They are most complicated of all in the temperate zones, where the Law of Gyration appears in its complete generality, and the storms, too, assume every known character. There is, however, one advantage which the temperate zones possess over the torrid, viz. that the violence of the cyclones which they experience is much mitigated from the fury exhibited by the same storms between the tropics, where they commit such fearful ravages.

Practical Rules.

IN the preceding pages we have sought to reduce to certain principles the behaviour of the vane and the barometer, under the influence of the different classes of storms which may occur in the temperate and torrid zones, and have indicated their variations, which are due to the Law of Gyration. We may now pass on to the solution of the following problem, viz. what are the indications which the instruments will give of the approach of a storm. Ships' barometers are certainly costly instruments, and are liable to breakage at the moment of danger*; but aneroid barometers can be got for a moderate price (50s. to 60s.), and may be fastened up, like a clock, to the cabin bulk-head. There is no doubt that, by a general use of these instruments, whose accuracy is quite sufficient for the purpose, danger might often be escaped, as the warning they give is so timely that it will often be possible to reach a safe harbour or to postpone putting to sea. The weather-scale (words), which is printed on most barometers, may lead to error; so that I shall give briefly the rules for the variation of the instrument.

1. REGION OF THE TRADE-WINDS.

The permanent wind of the torrid zone, NE. in the

---

* Thom recommends the sympiesometer which is described in the Edinburgh Journal of Science, No. 20, as particularly accurate in its indications.

northern, SE. in the southern, hemisphere, is divided by the intermediate zone of Calms into two parts. This zone does not fall exactly on the equator, but a little to the northward of it, so that the SE. Trade-wind extends as a S. wind a few degrees above the equator into the northern hemisphere. The breadth of this zone is greater in summer than in winter (*vide* Tables, pp. 20–24).

In the zone of the Trade-winds the level of the barometer varies very slightly in the yearly period; but on the outer edge of that zone stands about one or two tenths of an inch higher than on its inner edge; so that the barometer gradually falls to this extent as the ship approaches the equator. In the diurnal period it changes very regularly, reaching a maximum at about 9 A.M. and P.M., and a minimum after 3 A.M. and P.M. This variation, however, hardly extends above a tenth of an inch (p. 42).

In the centre of a cyclone the fall of the barometer is often an inch *; so that it follows at once that a sudden and rapid fall of the barometer indicates the approach of a cyclone.

The path of a hurricane divides that portion of the earth's surface, over which it is circling, into two regions, in

---

* In addition to the instances recorded above, we have the following notices of the fall of the barometer in certain localities : —

Ship *Duke of York*	at Kedgeree	in 1833,	2·70 in.
„ *Howqua*	„ Timor Sea	„ 1844,	2·20
„ *John O'Gaunt*	„ China Sea	„ 1846,	2·15
Brig *Freak*	„ Bay of Bengal	„ 1840,	2·05
Ship *Exmouth*	„ South Indian Ocean	„ 1846,	2·00
	Havannah	„ 1846,	1·95
Ship *London*	„ Bay of Bengal	„ 1832,	1·90
„ *Anna*	„ China Sea	„ 1846,	1·85
	The Mauritius	„ 1824,	1·70
Ship *Neptune*	„ China Sea	„ 1809,	1·55
Port Louis	„ The Mauritius	„ 1819,	1·50
Brig *Mary*	„ West Indies	„ 1837,	1·50

(Piddington, *Horn-Book*, p. 264.)

which the shift of the wind, as observed at a stationary point, is different. In the right hand * semicircle, on either hemisphere, it is ' with the hands of a watch' N., E., S., W., N.; in the left hand semicircle it is ' against the hands of a watch' N., W., S., E., N. This shows us that if a captain finds it necessary to lay his ship to, he must take great care to find out which is the proper tack. The right tack, in all cases, is that which will allow the ship, when the wind shifts, to come up with it. This would be the starboard tack in the right hand semicircle (starboard tacks taut; if the wind be ENE., course of ship N.), and the port tack in the left hand semicircle (port tacks taut; if the wind be NNE., course of ship E.). If the ship were on the wrong tack in the respective semicircles, she would have to fall off in order to follow the wind as it shifts, and meanwhile be in a very perilous position, unable to offer any resistance, as her yards would be taken aback, even though she were under bare poles.

If we go on and draw through the centre of the cyclone a line perpendicular to the direction of its advance, we shall divide the two semicircles into four quadrants. It is easily seen that the most dangerous quadrant of the two, which lie in front of the observer at the centre, must be that in which the wind is blowing towards the path of the storm. This will be the right hand quadrant on the northern, the left hand on the southern hemisphere. In these two quadrants a ship cannot get farther away from the centre by sailing with the wind free. There is then nothing else for the captain to do but to heave-to on the proper tack, viz. the starboard tack on the northern, the port tack on the southern,

---

* *Right hand* and *left hand* refer to an observer, placed at the centre of the cyclone circle, and looking forwards along the direction in which the storm is advancing.

hemisphere. If, however, the ship is in a locality in which the direction of the path of the cyclone is pretty certain, and finds herself, judging from the direction of the wind at the beginning of the storm, in such a position in the most dangerous quadrant, that the centre of the cyclone must pass very close to her, she will not run any greater risk by scudding before the wind and crossing the path of the cyclone, and then sailing out of the circle on the same course, than she would have run if she had been hove-to.

In the left hand quadrant of the northern hemisphere, the right hand of the southern, a ship can get farther away from the centre by sailing before the wind. If the hurricane is too violent to allow of scudding, or the ship is so far from the centre that she has nothing to fear, and would be taken too much out of her course by scudding, she may be placed on the port tack in the northern hemisphere and on the starboard tack in the southern.

The two other quadrants of the circle, those behind the centre, are the least dangerous quadrants, as the centre of the cyclone is moving, of itself, away from ships situated therein.

The cyclones of the Atlantic Ocean advance usually from SE. towards NW. within the torrid zone. Accordingly, if the wind shifts from ENE. through E. towards ESE., &c., the ship is lying on the right hand (i. e. NE.) side of the cyclone, and ought to steer towards the NE. in order to get away from the centre. This is impossible; so she must heave-to on the starboard tack, and lie successively towards N., NNE., NE., &c.

If the ship be hove-to, and finds that the wind is keeping in the N.E. point and increasing, and that the barometer is falling rapidly, the captain may be quite sure that he is close to the track of the centre of the cyclone, which is coming upon him from the SE. ; so that, after the lull, he may expect the hurricane to blow from the SW., and the barometer to rise rapidly. In such a case the ship ought to run before the wind towards SW., when she will soon find the barometer falling less rapidly, and the wind getting round towards N. ; she must, however, hold on for some time longer on her southerly course.

If the wind shifts 'against the sun' from NNE. through N. towards NNW., &c., the ship must be lying on the left hand side of the path of the cyclone in its advance from SE. towards NW., and should hold a SW. course. If this be from any reason impracticable, she must heave-to on the port tack, and lie successively towards E., ENE., NE., &c.

The following table, for the use of the seaman on the northern hemisphere, contains the position in which the ship should be laid, or the course she should hold, for every point from which the wind can possibly blow in the most dangerous quadrant ; it also contains the other quadrants, as is easily seen. Nos. 1 to 17 represent all the winds which could be especially dangerous to ships in any Typhoons, which are travelling in the directions between NE. to SW. and SE. to NW. ; Nos. 5 to 17 the West India hurricanes within the torrid zone ; Nos. 9 to 22 their course from the tropics to about the latitude of the Bermudas ; Nos. 15 to 25 their farther course in the temperate zone.

The 2nd column gives the horizontal tangents to the cyclone circle, or the direction of the wind at any point at the commencement of the storm. The 3rd column gives the bearing of the centre with regard to the ship. The 4th gives one series of directions in which the wind may shift, for each of which the 5th column gives the course to be held or the tack on which she must be hove-to. The 6th and 7th columns are related in a similar manner and refer to the most dangerous side of the cyclone.*

NORTHERN HEMISPHERE.

	Direction of Wind at beginning of Storm	Bearing of centre from Ship	Shift of Wind	Corresponding course to be held	Shift of Wind	Course to be adopted
1	NW.	NE.	NW. towards W.	SE.	NW. towards N.	
2	NW. by N.	NE. by E.	NW. by N. „ W.	SE. by S.	NW. by N. „ N.	
3	NNW.	ENE.	NNW. „ W.	SSE.	NNW. „ N.	
4	N. by W.	E. by N.	N. by W. „ W.	S. by E.	N. by W. „ N.	
5	N.	E.	N. „ W.	S.	N. „ E.	
6	N. by E.	E. by S.	N. by E. „ N.	S. by W.	N. by E. „ E.	
7	NNE.	ESE.	NNE. „ N.	SSW.	NNE. „ E.	
8	NE. by N.	SE. by E.	NE. by N. „ N.	SW. by S.	NE. by N. „ E.	
9	NE.	SE.	NE. „ N.	SW.	NE. „ E.	
10	NE. by E.	SE. by S.	NE. by E. „ N.	SW. by W.	NE. by E. „ E.	
11	ENE.	SSE.	ENE. „ N.	WSW.	ENE. „ E.	
12	E. by N.	S. by E.	E. by N. „ N.	W. by S.	E. by N. „ E.	
13	E.	S.	E. „ N.	W.	E. „ S.	
14	E. by S.	S. by W.	E. by S. „ E.	W. by N.	E. by S. „ S.	
15	ESE.	SSW.	ESE. „ E.	WNW.	ESE. „ S.	
16	SE. by E.	SW. by S.	SE. by E. „ E.	NW. by W.	SE. by E. „ S.	
17	SE.	SW.	SE. „ E.	NW.	SE. „ S.	
18	SE. by S.	SW. by W.	SE. by S. „ E.	NW. by N.	SE. by S. „ S.	
19	SSE.	WSW.	SSE. „ E.	NNW.	SSE. „ S.	
20	S. by E.	W. by S.	S. by E. „ E.	N. by W.	S. by E. „ S.	
21	S.	W.	S. „ E.	N.	S. „ W.	
22	S. by W.	W. by N.	S. by W. „ S.	N. by E.	S. by W. „ W.	
23	SSW.	WNW.	SSW. „ S.	NNE.	SSW. „ W.	
24	SW. by S.	NW. by W.	SW. by S. „ S.	NE. by N	SW. by S. „ W.	
25	SW.	NW.	SW. „ S.	NE.	SW. „ W.	

(vertical note between 4th and 5th columns:) Or else, heave the ship to on the port tack.

(vertical note in last column:) Ship to be hove-to on the starboard tack.

Cyclones appear to be less frequent in the south-east trade. If, in that region, the barometer is observed to be falling, while the wind is shifting from SSE. through S., SSW., SW., towards NW., the ship is on the north-west or right hand side of a cyclone which is advancing from

* (James Sedgwick.) The true Principles of the Law of Storms, practically arranged for both Hemispheres. London, 1854, 8vo.

Remarks on Revolving Storms, published by order of the Lords Commissioners of the Admiralty. London, 1853, 8vo.

NE. towards SW., and must therefore hold a NW. course, or else heave-to on the starboard tack. If the barometer continues falling while the wind holds in the SE., and is increasing, the ship is close to the path along which the centre is approaching. In this case the wind will chop

SOUTHERN HEMISPHERE.

	Direction of Wind at beginning of Storm	Bearing of centre from Ship	Shift of Wind	Correspond-ing course to be held		Shift of Wind	
1	S.	E.	S. towards W.	N.		S. towards E.	
2	S. by E.	E. by N.	S. by E. „ S.	N. by W.	Or else, heave the ship to on the starboard tack.	S. by E. „ E.	Ship to be hove-to on the port tack.
3	SSE.	ENE.	SSE. „ S.	NNW.		SSE. „ E.	
4	SE. by S.	NE. by E.	SE. by S. „ S.	NW. by N.		SE. by S. „ E.	
5	SE.	NE.	SE. „ S.	NW.		SE. „ E.	
6	SE. by E.	NE. by N.	SE. by E. „ S.	NW. by W.		SE. by E. „ E.	
7	ESE.	NNE.	ESE. „ S.	WNW.		ESE. „ E.	
8	E. by S.	N. by E.	E. by S. „ S.	W. by N.		E. by S. „ E.	
9	E.	N.	E. „ S.	W.		E. „ N.	
10	E. by N.	N. by W.	E. by N. „ E.	W. by S.		E. by N. „ N.	
11	ENE.	NNW.	ENE. „ E.	WSW.		ENE. „ N.	
12	NE. by E.	NW. by N.	NE. by E. „ E.	SW. by W.		NE. by E. „ N.	
13	NE.	NW.	NE. „ E.	SW.		NE. „ N.	
14	NE. by N.	NW. by W.	NE. by N. „ E.	SW. by S.		NE. by N. „ N.	
15	NNE.	WNW.	NNE. „ E.	SSW.		NNE. „ N.	
16	N. by E.	W. by N.	N. by E. „ E.	S. by W.		N. by E. „ N.	
17	N.	W.	N. „ E.	S.		N. „ W.	
18	N. by W.	W. by S.	N. by W. „ N.	S. by E.		N. by W. „ W.	
19	NNW.	WSW.	NNW. „ N.	SSE.		NNW. „ W.	
20	NW. by N.	SW. by W.	NW. by N. „ N.	SE. by S.		NW. by N. „ W.	
21	NW.	SW.	NW. „ N.	SE.		NW. „ W.	

about to NW. after the lull, and the barometer will begin to rise. In this case the ship must run before the wind towards NW., and she will soon find the wind shifting towards S., but must hold on, on a north-west course, for some time longer. If the wind shifts in the opposite direction, from E. through ENE., NE., &c., towards NW., the ship is on the south-east or left hand side of the path of the centre, which is advancing from NE. towards SW., and is therefore in the most dangerous quadrant,

u 2

so that all that she can do is to heave-to on the port
tack.

The table (p. 291) for the southern hemisphere corresponds to that just given for the northern.

There is one point which must be remembered in the
use of these tables, viz. that the shift of the wind within
the cyclone is that which would be observed at a stationary point. This may, in many cases, if the ship be
itself making considerable way, be converted into a shift
in a direction which is exactly opposite to that which
would have been observed if the ship had been stationary
in either of the cyclone semicircles, or at least may undergo considerable modification. This will always happen
if the ship be in the most dangerous quadrant, and be
scudding before the wind exactly as fast as the cyclone
is advancing, or perhaps be gaining on it. In order to
illustrate this, I shall take the case of a ship in the
middle of the most dangerous quadrant (enclosed between
NW. and NE.) of a cyclone which is advancing from
SE. towards NW. on the northern hemisphere. I shall
take the diameter of the cyclone at 300 miles, and its
rate of advance at 10 knots an hour. She would have
the wind E. at the commencement of the hurricane, and
this wind would shift towards S. in the course of the
storm if it were observed at a stationary point.

If the ship scuds ten knots an hour before the wind,
she will find that the wind shifts to NE. in the first twelve
hours and that the barometer falls rapidly. She will then
cross the path of the cyclone with this wind, and will
find that the wind shifts more rapidly half round the
compass, to SW., in the course of the next seven hours.
Between the fifteenth and sixteenth hour the barometer
will reach its lowest level, while the wind is NW., as the
ship will be only about thirty miles from the centre. The

ship will cross the path of the centre for the second time between the eighteenth and nineteenth hours, but this will be after the centre has passed on, and the storm will end with a S. wind after about thirty hours.

If we suppose the ship in the same place as before, but the rate of advance of the cyclone 1·5 times as great as the rate at which the ship is sailing, and therefore fifteen knots an hour, she will find the wind still holding in the E. and increasing during the first ten hours. In the next four hours the wind will have the normal shift towards SE., and be most violent at this last point, when the barometer is lowest. The cyclone will terminate after about twenty-eight hours, during the last ten of which the wind will have been due S. She will have approached to within ten or fifteen miles of the centre between the fourteenth and fifteenth hours.

A ship which is hove-to in a cyclone, though she may have one or two knots an hour drift, will always have a normal shift of wind, i.e. the shift which would be given by the preceding rules for either semicircle, the right hand or the left hand one. The seaman cannot be too strongly urged to heave-to in the first instance if he is in waters which are liable to these cyclones, and is led by the signs of the weather to expect one. If his ship is once hove-to, he may ascertain accurately his position with regard to the cyclone, and then, either hold on on his former course, or choose any other which may be advisable for him, or, lastly, heave-to on the proper tack.

As regards the period at which these storms are especially prevalent in the different parts of the torrid zone, we may say, as a general rule, that they usually occur when the sun is at its highest altitude, and reach their maximum of frequency at the end of this period; consequently, in September in the West Indies and the

China Seas ; in February and March at the Mauritius and in the South Indian Ocean; while in the Bay of Bengal they appear to be most frequent at the change of the Monsoons, i. e. in May and October. In addition to the figures already given, we find in Piddington's 'Horn-Book' (p. 296, third edition, 1860) the following table * : —

TABLE OF AVERAGE NUMBER OF CYCLONES IN DIFFERENT MONTHS OF THE YEAR, AND IN VARIOUS PARTS OF THE WORLD.

For what number of years ascertained	Locality	Authority	Jan.	Feb.	March	April	May	June	July	Aug.	Sept.	Oct.	Nov.	Dec.
123	West Indies	Nautical Magazine . .	...	...	...	...	...	1	2	13	10	7	...	...
59	,,	U. S. Journal, 1843, p. 3	...	...	...	...	...	1	5	13	13	9	...	...
300	,, ...	Royal Geographical Society . .	5	7	11	6	5	10	42	96	80	69	17	7
39	Southern Indian Ocean, 1809–1848	Reid, Thom, H. Piddington	9	13	10	8	4	...	...	...	1	1	4	3
24	Mauritius, 1820–1844	Mon. A. Labutte, Trans Royal Society of Mauritius, 1849 .	9	15	15	8	...	...	...	...	...	...	...	6
25	Bombay	. . . .	1	1	1	5	9	2	4	5	8	12	9	5
46	Bay of Bengal, 1800–1846	H. Piddington . . .	1	...	1	1	7	3	...	1	...	7	6	3
64	China Sea, 1780–1845	H. Piddington, Captain Kirsopp . .	...	...	...	...	1	2	5	5	18	10	6	...

The storms of the temperate zone in the southern hemisphere have more the character of gales than of cyclones, as is shown by the fact that the predominant

* I have quoted from the last edition of the Horn-Book, and the table is more complete than that which is given in Professor Dove's work. My reason for doing so is that I find the following remark on the table in the Horn-Book : —

'In the former editions of this work, I have stated that no cyclones have been known to occur in the month of May in the China Sea; but from a letter from Capt. E. T. F. Kirsopp, commanding the steamer *Juno* at Manilla, I learn that a severe cyclone was experienced in the Bay of Manilla and in the adjacent China Sea as early as May 4, 1850. . .
. . . . The essential matter, however, for us at present is the fact, that severe cyclones may occur in May in the China Sea, and thus upon the appearance of doubtful weather or an uneasy barometer the careful seaman will take due precaution.'
The importance of this fact will be, I hope, sufficient excuse for my adding it to the text. — *Trans.*

change of direction is in accordance with the Law of Gyration. In the '*Ondersoekingen met den Zeethermo-meter*' ('Experiments with the Marine Thermometer'), we read at p. 109—'The course of the winds in storms is, with few exceptions, from N. through W. towards SSW. Storms which begin with SSE. last longer, and do not change their direction much.'

The barometer stands lower during all storms in the southern hemisphere than when there is no storm, as is shown by the following table :—

Latitude . . .	35°—40°	40°—45°	45°—50°
N.	−0·498	−0·405	−0·490
NNE.	−0·498	−0·482	−0·291
NE.	−0·409	−0·441	−0·456
ENE.	−0·291	−0·449	− ...
E.	−0·350	−0·354	− ...
ESE.	−0·370	−0·295	−0·047
SE.	−0·272	−0·228	− ...
SSE.	−0·413	−0·378	−0·705
S.	−0·252	−0·478	−0·478
SSW.	−0·276	−0·394	−0·456
SW.	−0·315	−0·563	−0·311
WSW.	−0·323	−0·582	−0·378
W.	−0·425	−0·567	−0·478
WNW.	−0·374	−0·689	−0·571
NW.	−0·401	−0·547	−0·456
NNW.	−0·466	−0·433	−0·445
Mean . . .	−0·401	−0·414	−0·427

This shows us that the barometer stands lowest with the equatorial current, as the polar current does not produce so great a depression.

Ordinary tornados and thunderstorms occur on each hemisphere at the time at which the sun has his greatest altitude; so that they occur during our summer in the northern, during our winter in the southern, hemisphere. The hurricanes also assume the character of thunder-storms, i. e. they are accompanied by heavy rain and violent electrical explosions. The appearance of the sky, as a cyclone is coming on, is characterised by masses of clouds which are continually changing their form, and frequently by a bank of clouds, in the distance, of extraordinary blackness.

On dry land, in certain regions, the Trombs take the peculiar form of dust whirlwinds, of which Baddeley* has given a very vivid description. The amount of electricity which is excited by the friction of the sand is so great that he was able to obtain from an insulated wire, not only bright sparks, but a continuous discharge of electricity.

As to the motion of the waves in a cyclone, they move out from the centre in directions which are more nearly radii of the circle the farther they are distant from the centre. Consequently, they move from the centre towards the circumference in lines somewhat inclined in the direction in which the storm is rotating. Reid has already drawn special attention to this fact.

Hence we obtain the following differences in this respect between the three classes of storms which we have considered.

1. In a cyclone, the waves move at right angles to the direction of the wind, and the more so the farther they are from the centre.

2. In a heavy gale, they move in the direction of the wind.

3. In the case of a wind which has been stopped by another, they move against the wind. (Seamen then say that two winds are fighting with each other.)

All that we have said hitherto has referred to the torrid zone, properly so called, not to the outer edge of the

* Whirlwinds and Dust-Storms of India, illustrated by numerous Diagrams and Sketches from Nature.

Trade-winds, which shifts up and down with the sun like the belt of Calms. Hence the torrid zone will be bounded in all parts, excepting where the Monsoons are prevalent, by a belt, in which calms are of frequent occurrence, and which is called the sub-tropical zone. This zone forms a complete contrast to that which lies close to the equator. At the latter the air is continually ascending, at the former it is descending. At the equator the barometer is low, at the tropics it is high. At the equator the rain falls in summer, in the sub-tropical zone in winter. At the equator the two winds flow towards each other, in the sub-tropical zone they flow in opposite directions away from the tropics. All stations which belong to the sub-tropical zone are included during the summer in the Trade-wind, on its prolongation backward at that season, and during the winter are outside its area. Such a prolongation in the direction of the poles as that described, is manifested in its most extensive form in the case where a great desert, like that in Northern Africa, prolongs the torrid zone into higher latitudes to a disproportionate extent. Hence, during the summer, northerly winds are prevalent in the Mediterranean, and are known under the name of *Tramontane*, while the *Sirocco* attains as exclusive a predominance in the winter, being the return counter Trade which has descended to the surface of the earth. This is the reason that the seaman finds, at the commencement of the Trade-wind, a more northerly wind on the east side of the Atlantic Ocean than on its west side.

2. District of the Monsoons.

During the summer months in the Indian Ocean, the SE. Trade-wind is prevalent in that part of the torrid zone which is in the southern hemisphere, and the SW. Monsoon in that part which is in the northern.

In the winter months the NE. Trade-wind is felt to the north of the equator, and the NW. Monsoon to the south of it.

In consequence of this alternation, the NE. Trade is termed NE. Monsoon, and the SE. Trade SE. Monsoon.

In spring and autumn, in the so-called 'change months,' there are calms; on the coast, sea and land winds in the daily period. The change from one Monsoon to the other is usually accompanied by a storm, 'the breaking-up of the Monsoon.'

The SW. Monsoon extends much farther to the N. (lat. 30°) in the northern hemisphere than the NW. Monsoon towards the S. in the southern; however, the latter extends also far to the S. on the African coast.

The rainy season is observed, as it is in the region of the Trade-winds, when the sun is highest. It is, therefore, during the SW. Monsoon on the northern, and during the NW. Monsoon on the southern, hemisphere.

There is, however, a difference in the behaviour of the barometer. In the region of the Trade-winds it remains nearly constant throughout the year, in that of the Monsoons it varies regularly. During the SW. Monsoon the barometer stands several tenths of an inch lower than in winter, especially in the northern portion of the district; and similarly in the southern hemisphere, it stands lower during the NW. than during the SE. Monsoon. At the equator this annual variation of the barometer disappears nearly entirely, in consequence of the compensation of the two opposite variations.

The rotations of the storms of the China and Indian Seas are the same as those in the same latitudes in the region of the Trade-winds; however, on the coast of

China, their motion is rather from E. to W. than from SE. to NW. One important distinction is to be observed, that, in the district of the Monsoons, the storms are felt with great violence in the southern hemisphere as well as in the northern.

The rotation of the vane during the Typhoons, although the direction of the rotation of the air itself is quite fixed (contrary to the hands of a watch), is yet, in consequence of the uncertainty of the direction in which the centre may possibly move, less fixed than during the West Indian storms. They take place during the SW. Monsoon, and occur up to November, being most frequent in September.

If the Typhoon move from NE. to SW., on the NW. side of its path, the rotation is N., NE., E., or with the sun; on the SE. side WNW., SW., SSE., or against the sun. On the south coast of China the rotation is generally N., NE., E., SE., as the Typhoons which are passing from E. to W. usually pass to the south of the coast.

According to Thom, the storms are never felt in the Indian Ocean excepting at the time when the NW. Monsoon is prevalent between the equator and the tenth or twelfth degree of south latitude, and are most common just after the winter solstice, when the sun is turning back from the Tropic of Capricorn.

The cyclones are most common in the district between the SE. Trade-wind and the NW. Monsoon, which is called the region of the 'variables.'

The rotatory motion takes place in the direction from E. through S. towards W. and N. The intensity of the cyclone increases regularly towards its centre. At the centre itself there is a dead calm, and the greatest violence of the storm is experienced at the edge of this calm circle. The diameter of this circle is greatest when the storm is just commencing. If the rotatory motion increases in violence, the diameter of this circle is decreased to about ten or twelve English miles.

The advance of the cyclone, up to lat. 20° S., is at the rate of 200 to 220 miles in the twenty-four hours. From that point it becomes less rapid up to the outer edge of the SE. Trade.

The direction of the advance is from lat. 10° S. near the Indian Archipelago, to lat. 28° or 30° on the east coast of Africa ; first towards WSW., then towards SW. by S., and lastly towards SSW.

Throughout the whole of the cyclone torrents of rain fall, which are more violent in front of it than behind it. The clouds are dark, massive, and lead-coloured, as the centre is approaching. Electrical explosions are most frequent on that side of the cyclone which is nearest to the equator.

The sea is disturbed irregularly to the distance of 300 or 400 miles during every such storm.

The barometer falls rapidly as the centre of the cyclone approaches, but the lowest level appears to occur a little before it passes.

3. NORTH TEMPERATE ZONE.

In my ' Meteorological Investigations,' 1837, I have stated expressly that the principal characteristic of the climate of the temperate zone is the alternation of two currents of air, of which the one flows from the pole to the equator, the other from the equator to the pole ; and in my ' Non-Periodical Variations of Temperature on the Surface of the Earth,' six parts, 1840–59, as well as in the ' Representation of the Phenomena of Temperature by Means of Five-Day Means,' I have proved that these currents move simultaneously in proximity to each other.

Hence, in these localities we can no longer speak of a
constant direction of the wind, as in the zone of the
Trade-winds, nor of one which changes periodically, as in
that of the Monsoons, but only of a mean direction. This
mean direction is nearly SW. * in the north, and NW. in
the south, temperate zone ; inasmuch as the equatorial currents are more prevalent than the polar. In Europe this
westerly direction is more southerly in winter than in
summer; in America the reverse is the case; and the
gradual transition from one of these conditions to the
other takes place on the Atlantic Ocean. Violent storms
are felt here less frequently in summer than in the winter
months, and in the Mediterranean Sea at the transition
of one season into the other; whence they are called here
' equinoctial storms.' These storms are either, 1, gales,
i. e. ordinary winds whose intensity has been greatly
increased, and which cause the vane to rotate with the
sun, but through comparatively small arcs; 2, cyclones
from the torrid zone, which have changed their path on
crossing the outer limit of that zone, and have taken a
course from SW. to NE. in the northern, and one from
NW. to SE. in the southern, temperate zone ; 3, currents,
which, by their mutual interference, have checked and
then repelled each other; or 4, storms, produced by the
sudden intrusion of the cold polar current into the warm
equatorial current, a case of which many remarkable instances have been noted. Hence, the barometer during
the yearly period neither remains steady nor varies
regularly, but is subject to oscillations, which are greater
in winter than in summer. The mutual alternation of
the currents is indicated by a rotation of the vane with
the sun, i. e. S., W., N., E., S. on the northern, and
N., W., S., E., N. on the southern, hemisphere. Hence we

* See Note, p. 82.

derive the following general rules for the variations of the meteorological instruments : —

Since the southern current is warm, moist, and rarefied, the northern, on the contrary, cold, dry and dense, we derive the following rules for their alternation; and we must remember that the cold polar current appears first in the lower strata of the atmosphere, while the warmer equatorial current will always have existed for some time in the upper strata before it is felt below. The changes of weather on the west side of the compass are, therefore, simultaneous with the changes of the barometer, while on the east side the indications of the barometer always precede the fall of rain which takes place. If the wind shifts from S. to N. through W., the barometer rises and the thermometer falls. This transition is characterised in winter by heavy falls of snow, in spring by sleet showers, and in summer by thunderstorms, after which the air becomes much cooler. If the wind veers from N. to NE., we have clear weather, the air is dry, the barometer high, and in winter intense cold follows, with great clearness of the atmosphere. As soon as the barometer begins to fall the wind gets round to the E. ; the sky, previously deep blue, covers itself with thin whitish clouds, and the snow which falls comes from the S. wind, which has already set in above. If the barometer falls rapidly, the snow turns to rain, and a thaw sets in, when the wind veers farther through SE. and S. towards SW.*

The transition from a clear sky to an overcast one

---

* The first notice of this transition is to be found in Drebbel, *De Naturâ Elementarum* (concerning the nature of the elements), 1621. ‘If we see a thick cloud rising in summer not far from the south-west, we expect, and also find, that a SW. wind will soon blow, then a W., NW., and lastly a NNE. You see also why the E. or SE. wind brings such heavy and continued rain with it in Holland and the adjacent countries. I could very easily explain, on natural grounds, the reasons of all these phenomena.'

usually commences with the appearance of fine streaky cirrus clouds, which gradually change to cirrostratus, and then the uniform coating of cloud is complete. This cirrus represents the equatorial current, seen from beneath, which has already set in above, and marks its progress by the streaks of cloud. The water, on its condensation from the state of vapour, assumes the solid form at once; so that these high clouds are not composed of bubble steam *, but of minute spiculæ of ice, and they give rise to the halos of the sun and moon, which are caused by refraction of the light, and to the so-called rings, mock-suns, and mock-moons.

If these appearances accompany a falling barometer, it is a sure sign that wet weather is coming on.† The reason that the long streaks of the cirrus appear to us as arcs of circles which diverge from one point of the horizon and reunite at the opposite point, is that they are projected on the apparently curved surface of the sky. This apparent curvature of the cirrus differs from the lateral feathery off-shoots of the same clouds, which show that the direction of the upper wind is not quite constant. This latter form of cirrus, consequently, is a less certain sign of rain than the long arched clouds are. There is another form of cirrus, which does not always indicate rain, as the air gets warmed during the day and ascends; if the temperature be high this ascending current sometimes reaches as high as the cirrus above, and then the latter

---

* *Nebelbläschen* is the word translated 'bubble steam' (*vapeur vesiculaire*, Fr.). It indicates visible steam, as distinguished from vapour, which is invisible. (*Trans.*)

 † The hollow winds begin to blow,
 The clouds look black, the glass is low;
 Last night the sun went pale to bed,
 The moon in halos hid her head:
 'T will surely rain.

breaks up into small cumuli, which are known in Germany under the names of *Schäfchen*, *Lämmer-Gewölk* in South Germany, *brebis* in France, *fleecy clouds* in England, and were called by the Romans *vellera lanæ*. Howard calls them 'cirro-cumuli.' In the south of Europe they are said to be a sign of rain. In Northern Germany this is not the case, according to my observations.

When the atmosphere is warm and dry, the outlines of distant objects become indistinct and hazy, owing to the dust which is suspended in the air, and the sun appears reddish. If easterly and northerly winds have lasted for a long time in summer, with very dry weather, and a moist wind sets in, its aqueous vapour condenses itself at once on the dust which is in the air, which thus becomes heavy and sinks to the ground. Under these circumstances, the air becomes very clear, and in mountainous countries the mountains appear quite close, and the waterfalls are heard more distinctly. This is considered an infallible sign of rain.

The rain comes, as a general rule, from the west side; so that a clear sunset is a proof that there is no rain coming from that quarter for some time. Hence, this is considered to be a sign of fine weather.

In the evening, when the air ceases to ascend, the clouds sink, and are dissolved in the warm strata below. From this nothing can be augured for the following day. There is an old French proverb :—

> Temps, qui se fait beau la nuit,
> Dure peu quand le jour luit.*

If the atmosphere is very damp, evaporation cannot go on, and this feeling produces in us the sensation which we designate by sultriness, *drückende Luft*. The direct

---

* Weather which clears up at night, will not last when the day breaks.

action of the sun is then more felt, and we say the sun is scorching.

If the south wind sets in suddenly in the upper strata in winter, the rain falls at once at that level, and small transparent grains of ice, i. e. rain frozen while falling, reach the ground. We say then that *Glatteis* (glazed frost) is falling, as the rain which soon sets in freezes on the ground and glazes it. We may then expect a SW. storm with a great fall of the barometer.

In winter, rain, with a west wind and a rising barometer, turns to snow ; snow, with an east wind and a falling barometer, to rain.

In spring, if the wind shifts through W. to N., we may expect the weather to clear up suddenly and night frosts to set in, even though the thermometer, at a little height above the ground, may not fall below the freezing point.

Heavy thunderstorms, which come up with an E. wind while the barometer is falling, do not cool the air. We say that it is still sultry and there will be another thunderstorm. The air does not grow cooler till a thunderstorm comes up from the W. and the barometer begins to rise.

If several thunderstorms come on in succession from the W., each successive storm is usually more northerly than that which has preceded it.

In the case of thunderstorms from the W., the under current is usually more northerly than the upper one; consequently the true thunderclouds (cirro-strati) drive more or less at right angles to the lines of the cirrus above them.

The greater the difference of temperature between the two currents which displace each other, and accordingly the greater contrast there is between their directions, the more likely they are to produce a thundercloud. The winds stop each other's path and produce a calm before

x

the storm comes up. The cold wind then breaks in suddenly, and it is a mistake to say that the thunderstorm has made the wind change.

The winter thunderstorms in Norway are westerly storms, in which the under current is shifting quickly towards N. The barometer rises and cold follows. They are preceded by a thaw, mild weather, heavy rain, and southerly winds.

Our winter thunderstorms in Germany, which are rare, exhibit the same character. There is, however, another type of these storms. This is exhibited when the equatorial current sets in with great violence. In such a case, the thunder and lightning are often so tremendous that we say the sky is bursting open (*der Himmel öffnet sich*). They are followed by a complete spring wind.

The time at which thunderstorms are most common is regulated by the commencement of the rainy season. They are most common in the height of summer in the torrid zone; in mid-winter in the district of the sub-tropical rains at the outer edge of that zone; in spring and autumn in the south of Europe; and in the middle of summer in that part which is north of the Alps, with the exception of Norway. They are, on the whole, rare in the frigid zone, but yet do occur there up to high latitudes. Lastly, they occur in volcanic districts, as secondary results of the rapidly ascending current above a volcano in eruption, and at times at which they are never observed unless under these circumstances.

If bad weather continues for a long time, the vane oscillates between SW. and W., and the barometer fluctuates slightly. This is the true equatorial current.

Thunderstorms in spring lie at a low level, and do not last long; they are usually followed by a return of cold weather. They are at times accompanied by sleet or snow, and they frequently do considerable damage by means of lightning. The lower wool-pack clouds drive with a WNW. wind, the upper cirrus with SSW.

If the barometer rises very quickly, this indicates, not that the southern and northern currents are interfering with each other laterally, but that they have met and mutually stopped each other's way. A severe storm is sure to follow; and if the barometer falls as quickly as it has risen, it shows that the southern current has prevailed, and that the danger is therefore close at hand. In this case, the lettering of a barometer which bears 'very dry' for this level is totally wrong.

If in winter a cold and a warm wind meet each other, and the southerly current has not sufficient force to over-come the resistance of the northerly current which opposes it, the barometer rises to a great height at the line of contact, and a thick fog appears there. This fog often disappears suddenly and reappears again, according as the southerly current gives way a little, and the place of observation comes off the line of contact into the true northerly current, and vice versâ. If severe cold follow such a fog, it shows that the northerly current has finally prevailed.

If the barometer at any place oscillates violently, and the air remains at rest there, the disturbance must lie in a lateral direction. At times in winter the southerly current prevails over a large area, and the barometer is low, the air delightfully mild. Under these circumstances there is a very severe winter, with a high barometer some-where in the neighbourhood. It is possible that this cold air may force its way, as a storm, into the warm and rarefied air in its neighbourhood, and cause the barometer to rise rapidly.

On the Atlantic Ocean, if the wind veers against the

x 2

sun, i. e. from NE. through N. to NW., and the barometer
be falling rapidly, the ship is probably in a cyclone whose
centre lies to the SE. and is moving towards the NE. In this
case she must steer NW., in order, if possible, to get away
from the centre of the cyclone, where the danger is
greatest. If it veer from SE. to S. and SW., and the
barometer be falling, the ship is either in an ordinary
gale, or in a cyclone whose centre lies to the NW. In
the latter case she must steer SE. ; and this course is the
best for her to take in general, as the ordinary SW. gales
usually increase in intensity to the westward.

If the wind continues a gale from the SE., and the
barometer keep falling, it is probable that the ship is
exactly in the path of a cyclone which is moving from
SW. to NE. If the barometer still fall, and the wind
keep in the same quarter but increase in violence, the
centre is coming closer. If the ship come into the
centre of the cyclone, there is a sudden lull when the
barometer is at its lowest level. This is the moment of
greatest danger, as the storm will recommence from the
quarter diametrically opposite, viz. from NW. In these
cases the vane gives the tangents to the cyclone. In the
West Indies these storms travel from SE. to NW., and the
vane therefore shows NE. before the centre has passed
and SW. afterwards. As soon as they reach the boundary
of the torrid zone they turn at a right angle, and travel
from SW. to NE. It is only this portion of the cyclone,
on its altered course, which we feel in Europe, and, owing
to the increase of its diameter, we are not exposed to its
fury until this has been diminished. In these instances
a fall of the barometer indicates the increase, a rise the
decrease, of danger.

Cyclones of a smaller diameter, known as ‘Trombs,’ at
times do great damage in our forests : their lateral extent

is comparatively small; however, in the neighbourhood of the path of their centre they are capable of blowing down trees,* unroofing houses, and lifting heavy articles from the ground. In the progress of such a whirlwind its axis frequently receives a considerable inclination in the way it is moving, in consequence of the resistance presented to the motion of the air by its friction with the ground. It is probable that many of our thunder and hailstorms are to be attributed to this circumstance. The grain of sleet, first formed at a great height in the air, makes several revolutions in the inclined whirlwind, and in its passage through cold and hot strata alternately it obtains the shell of ice, which covers the grain of sleet, like a grain of snow, in the centre, and at last becomes so heavy that it falls to the earth. The characteristic noise which precedes a hailstorm is owing to the rotatory motion of the hailstones before they fall. Such hailstorms, and many severe thunderstorms, present the striking appearance of a long, almost horizontal, column of clouds which is rolling on, and when projected on the sky appears more or less bent. At times the dark bank of clouds covers itself with a number of brighter stripes of grayish clouds, which envelope it, like a waterfall does the cliff over which it falls. The edges of the whirlwind seem to favour the formation of hail, in consequence of the fact that there the circles described by the hailstones are largest, and consequently the difference of temperatures which they have to pass through is greatest. It has been very often observed that the district where hail fell,

---

* In September 1848, I saw what such a Tromb had done in the forest of Biesenthal near Neustadt-Eberswalde. The track was like a long trough with sloping sides. Along the centre the trees had all been broken off close to the ground, and towards the sides you found them broken off nearer and nearer the top, and many of them twisted together.

whose breadth is never great, has been double, with a district in the middle where it has only rained. The reference of the formation of hail to the whirlwind explains the fact that the boundaries of the hail district are often very clearly marked. The barometer is not much affected by hailstorms : they are local phenomena which it cannot indicate, as it measures the total pressure of the atmosphere, and therefore only gives notice of phenomena which are on a great scale.

The sudden squalls which accompany these thunderstorms are sometimes very dangerous to ships if their upper sails are not reefed. In the year 1850, at Heringsdorf, I witnessed a thunderstorm like this, on a day which was otherwise very fine. It lasted a very short time, and there was only one clap of thunder, like a cannon shot. On the passage to Rügen next day, I saw a ship at the mouth of the harbour of Swinemunde, which had capsized in bright sunshine, so suddenly that the corpse of one of the crew could not be got out of the cabin.

Lettering on barometers loses its value from the fact, that the difference of temperature, and therefore of density between the two currents, is much greater in winter than in summer. Inasmuch as the fluctuations of the barometer in winter are much greater than in summer, it is evident that the scale for winter should have at least double the extent that it has for summer. It is easy to see how such lettering has arisen. Correctly speaking, the highest mark should be NE. wind, or, better, 'polar current;' that in the middle E. or W. wind, or, better, 'transition;' the lowest SW. wind, or, better, 'equatorial current.'. The air of the polar current flows from a colder to a warmer climate; so that, as its capacity for the absorption of aqueous vapour is increasing, the effect of this increase is entered on the scale as 'very dry.' At the

transition of one current into the other, rain falls, owing
to the mixture of the air belonging to the two currents;
but at the same time the weather either breaks or clears
up for a time, so that this point is marked ' change.' The
southern current, as it moves into higher latitudes and
comes into contact with a surface whose temperature is
continually decreasing, discharges the aqueous vapour
which it has absorbed; so that we have at the point
corresponding to it ' much rain.' If the southern current
forces its way to the northward very rapidly, the contrast
between the pressure exerted by the rarefied air of which
it consists, still farther diminished by the condensation of
its aqueous vapour, and the mean value of the atmospherical pressure, is greatest, and consequently the lowest
mark on the barometer is ' stormy.'

From what has been said above, it is easy to see that,
as the barometer rises with rain on the W. side, and falls
with it on the E. side, of the compass, it is impossible to
lay down rules for the weather which do not take the
direction of the wind into consideration, as has been
attempted by many persons. At times the phenomena of
the one side pass into those of the other, without any
change or interruption in the form of the discharge
having taken place. If after severe cold it begins to
snow, and the vane moves from E. to SE., the barometer
falls and the cold grows less intense; however, it does not
always rise above the freezing point. In this case, when
the wind gets round to S., the snow does not turn to rain,
and if this S. wind is, in a short time, in its turn displaced,
the fall of snow is uninterrupted, but really it consists of
two separate formations : one, while the barometer is falling, in consequence of the displacement of a colder by a
warmer wind; and the other, while it is rising, from the
converse change. The rule that fresh snow brings fresh

cold arises from the fact, that snow is more usual with W. than with E. winds. It is also easily seen that snow can never fall when the temperature is very low, since it arises from the contact of two currents whose temperature is different. It is certainly true that some snow falls when the cold is very intense ; but in this case it does not take the form of flakes, but rather that of spiculæ of ice, which owe their origin to a stratum of clouds which belongs to a warmer current, lying at a great height in the atmosphere. These needles, passing in their fall through very dry air, cannot increase in size, and hence cannot assume the form of flakes. If the variations of the barometer in summer and winter were of equal extent, or, in other words, were the difference of pressure of the two currents constant, the barometer would, generally speaking, stand lowest during rain. This is, however, not the case as regards the yearly mean ; for the depression of the barometer during S. winds below its mean level is greater in winter than in summer, while the usual form of the discharge in winter is that of snow. During the course of one single revolution of the vane, the barometer is lower in rain than in snow.

If in spring, in the centre of Europe, the barometer be high, and the wind easterly, we may expect to have strong S. winds, accompanied by heavy rain, in the south, e. g. in the Mediterranean ; since the high level of the barometer is owing to the fact, that the wind blowing from higher latitudes has been prevented from finding a passage to the south by the sirocco, which is blowing in the opposite direction, and which is the upper current returning to the surface of the earth at the outer boundary of the Trade-wind. We further deduce the following fact from the Law of Gyration :—Southerly winds in high latitudes are more westerly ; northerly, in low latitudes more easterly: the only winds which can preserve their direction unaltered over a large area are due E. and W. winds.

Rotations of the vane against the sun, which extend beyond S. or E., indicate cyclones; if they only extend from NW. to SW., or from ENE. to NNE., they are often only a return of the vane to its original position, indicating, in the one case, that the equatorial, in the other, that the polar, current continues prevalent.

4. SOUTH TEMPERATE ZONE.

The regular rotation of the wind in this zone is with the sun, i. e. from S. through E. and N. to W. and S.; and similarly in cyclones, when the ship is on the NE. side of the path which is travelling from NW. to SE., it is from NNE. through N. and NW. to W. and WSW.; and in both cases the barometer falls until the wind reaches NW., and then rises. The only difference is that in a cyclone the temperature does not vary to any extent, while in the case of the alternation of the regular currents it rises while the barometer falls, and vice versâ. If the wind veer from W. through SW. to SE., the barometer in general rises and the thermometer falls. The prevalent wind, when the barometer is highest and the thermometer lowest, is SE., and when the converse conditions are fulfilled, NW., especially in the cold season; the atmosphere also is clear with SE., and thick with NW. If the wind veer from W. through SW. to SE., the weather clears up; if from SE. through NE. and N. to NW., it breaks, and there is rain. If the wind veer against the sun from ENE. through SE. to S., the ship is probably on the SW. side of the path of a cyclone travelling from NW. to SE. The rules for finding the position of the centre and the course to hold have been already given.

The only parts of the earth which I have not considered are the frigid zones. The stormy periods here seem to be the summer, and the transition from winter to summer ; the winter itself is, comparatively speaking, a time of calms. In the American polar sea, the barometer stands at its highest level in spring. According to Ross's observations, the permanently low level of the barometer, which was first observed at Cape Horn by Krusenstern, appears to extend far into the antarctic zone. This district of barometrical depression is of far greater extent than that in the vicinity of Iceland. The cold air, lying over the ice-fields, seems to stop the most violent south winds, which accordingly deposit their aqueous vapour in heavy falls of snow in their attempts to force a passage ; hence large floes which are rotating are surrounded by a wall of snow. Very little of this snow penetrates to the interior of the floe, while the different points of the circumference come successively into the area where the contest is going on (Scoresby). In the immediate neighbourhood of the pole the rotation of the vane becomes complicated ; since the influence of the rotation of the earth on the wind changes as soon as storms pass the pole, from the fact that the velocity of rotation of the surface with which the air comes in contact, which had been previously decreasing, begins to increase again. The dense fogs which arise from the difference of temperature of the sea, and the very cold air lying immediately above it, and the similar fogs which are due to the difference of temperature of the air over the sea and over the ice-floes, are the prevailing form which the condensation of moisture takes in these regions. This form is also not unusual in spring in the N. Atlantic Ocean, owing to masses of ice which are drifting southwards, and is an indication of the proximity of icebergs, especially off Newfoundland.

There are too few data for the Pacific Ocean to permit me to enter into a detailed examination of the differences between the phenomena on that and on the Atlantic Ocean. The description of the storm in the harbour of Avarua, in Raratonga, which is given by Williams,* shows that the hurricanes are very violent in those seas. He says : 'The whole island quivered to its centre when the waves broke on the coasts. A vessel belonging to the missionaries was carried over a marsh into a wood of large chestnut trees some hundred yards from the shore. The rain fell in torrents from morning to night.'

The practical rules which have just been given are intended to serve a twofold purpose. Firstly, they will indicate to the seaman the conclusions as to approaching weather, which he may draw from the appearance of the sky and the behaviour of the meteorological instruments, especially of the barometer. Secondly, they will show him which of the phenomena are, as yet, imperfectly explained, and which, therefore, demand a more accurate investigation by means of additional observations. It is very satisfactory to find that practical seamen like Fitzroy, Maury, Van Gogh, Andrau, and Jansen, are taking steps to indicate to the officers of the naval and mercantile service what points it is important, not only for science, but also for themselves, that they should ascertain. By this means central stations, like the Board of Trade, the National Observatory at Washington, and the Meteoro-

* Narrative of Missionary Enterprises in the South Sea Islands.

logical Institute of the Netherlands, have been established, at which the fragmentary materials are registered and worked up into a collected whole.

In my opinion the strictly meteorological element of such investigations has not been brought forward in a sufficiently prominent manner in the instructions furnished by these institutions; and I have, therefore, sought to supply the deficiency by the present work.

The theory which has been here propounded, and which has been developed by me in various treatises since the year 1827, assigns answers to the following problems :—

1. Why the storms of the torrid zone appear more frequently in certain districts than in others.

2. Why they take the form of cyclones; and why the rotation in a cyclone is in a different direction on the northern to what it is on the southern hemisphere.

3. Why they move in fixed directions within the tropics, and turn at a right angle as soon as they cross the boundary of the torrid zone.

4. Why the cyclone increases in diameter and decreases in intensity when this change of path has taken place.

5. Why the form of the storms of the temperate zones presents more variety,* and why in these districts certain

---

* In what cases these are not to be distinguished (from a local point of view) has already been explained. The following is an important historical example. Macaulay, *History of England*, vol. ii. p. 455, says :— 'The weather had indeed served the Protestant cause so well, that some men of more piety than judgement fully believed the ordinary laws of nature to have been suspended for the preservation of the liberty and religion of England. Exactly a hundred years before (they said) the Armada, invincible by man, had been scattered by the wrath of God: civil freedom and Divine truth were again in jeopardy, and again the obedient elements had fought for the good cause. The wind had blown strong from the east while the prince had wished to sail down the Channel, had turned to the south when he wished to enter Torbay, had sunk to a calm during the disembark-

characters of storms are more prevalent at certain seasons and at certain localities than at others.

In conclusion, I wish to draw attention to the fact, that the theory which has been developed in the foregoing pages is only intended to exhibit the principles from which we may deduce, not only the Trade-winds and Monsoons, but also the regular movements of the atmosphere of the regions of changeable winds, when it is not disturbed by storms. According to the explanation which has been given of them, the hurricanes tend to accelerate the earth in its motion round its axis, whereas the constant Trade-wind tends to retard it. The compensating element for the conservation of the earth's rotation, which would otherwise be affected by the great Trade-wind currents, is furnished by the various westerly currents, viz. the predominant equatorial, and consequently westerly current in the temperate zone ; the SW. Monsoon of the northern, and the NW. Monsoon of the southern, Indian Ocean ; and, lastly, the hurricanes. Despite the destructive action of the last-named movements, they must still be regarded, in the general sense of the term, as agents in the conservation of the vital force in the great organism of the earth. The whole of the phenomena which we have described furnish us with confirmatory proofs, on a great scale, and borrowed from the earth itself, of the great fact that the earth rotates about its axis—a fact whose first discovery is due to the science of astronomy.

In this second edition I have discussed the theories of others at greater length than I did in the first. My reason for doing so has not been to bring forward my own theories more prominently, but to show the error of the idea that all atmospherical phenomena may

ation, and as soon as the disembarkation was completed had risen to a storm, and had met the pursuers in the face.'

be discussed according to any one cut-and-dry pattern. There are so many agencies always at work disturbing the equilibrium of the atmosphere—the radiation, whose extent varies from day to day—the infinite variety in the surface of the ground—the ocean currents, and the different forms in which aqueous vapour presents itself — that the Calms ought to excite our astonishment in a much higher degree than the Wind. The atmosphere is eternally striving to attain equilibrium without ever succeeding. The character of the disturbance itself, and the process of restoration of the equilibrium, exhibits in each case a distinct type; so that the problem which presents itself to the meteorologist is to discover the typical form of the phenomenon, which presents in each several case of its occurrence variations of more or less extent from the original type. Generally speaking, the principal types have been distinguished as *Wirbelwind* and *stetiger Sturm*, *hurricane* and *gale*, *ouragan* and *tempéte*; but these two forms pass into each other by such insensible gradations, that a gale may become a whirlwind at one point of its course, without being a true cyclone in the strict sense of the word. I have been anxious to show, as clearly as possible, how unjustifiable it is to confound the effects of the Law of Gyration with those of rotatory storms—an error into which many have fallen, and which has not yet disappeared. The reason that many physicists refuse to recognise the existence of cyclones has arisen from the fact, that the disciples of the cyclone theory have thought that they had discovered a cyclone wherever the rotation of the earth about its axis exhibited itself in the rotation of the vane,* and have consequently left blots which

* The influence exerted by the motion of the earth is well described by Herschel in the following words (Meteorology, p. 57): — 'To form a right estimate of its importance, it is only necessary to observe, that of all the

their adversaries were not slow to hit. Those who, in exposing such errors as these, willfully shut their ears to Nature when she speaks to them in such unmistakable language as she does in the Typhoons and West India Hurricanes, ignore the problem which it is the business of meteorologists to attempt to solve, viz. to interpret her language, however varied the expressions which she employs may be.

winds which occur over the whole earth, one-half at least, more probably two-thirds, of the average momentum is nothing else than force given out by the globe in its rotation in the Trade currents, and in the act of reabsorption or resumption by it from the anti-Trades.'

INDEX.

AFR

A FRICA, Coast-winds of, 67.
  Air, dry, annual variation of pressure of, in Asia, 54.
— — law of variation of pressure of, 138.
America (South), Coast-winds of, 69.
Anemometrical Tables: Northern Hemisphere, 96–107.
— — Ireland, West of, 82.
— — Melbourne, 115.
— — Monsoon Region, 48.
— — Trade-wind Region, 26–35.
Aqueous Vapour, annual variation of tension of, in Asia, 52.
— — law of variation of tension of, 137.
Aristotle, notice of the winds, 77.
— — — Law of Gyration, 88.
Arys, Barometrical Table for, 124.
— — — — — monthly, 133.
Ashes, fall of, in Trade-wind Zone, 36.
Asia, curves of rarefaction in, 61.
— extent of rarefaction in, 62.
Atmosphere, pressure of. *See* Barometer.
— movements of, 270.
— — — Maury's theory of, 271.
Atmospherical waves, 266.
Australia, effect of, on NE. Trade-wind, 72.

B ACON, Lord, notice of Law of Gyration, 88.
Ballot, Buys, on Law of Gyration, 98.
Barbadoes Hurricane (1831), description of, 205.
Barometer, annual variation of, in Central Asia, 61.
— — — — — Africa, Northern, 63.
— — — — — S. Hemisphere, 71.
— — — — — Trade-wind Region, 42.
— laws of variation of, N. Hemisphere, 123.

CYC

Barometer, laws of variation of, S. Hemisphere, 134.
— fall of, in Cyclones, 158, 202, 286.
— height of, in Trade-wind Region, 187.
— rise of, may precede a storm, 307.
Barometrical Windrose, S. Hemisphere, 295.
Berlin, anemometrical results for, 97.
Bermuda, anemometrical results for, 105.
Bombay, anemometrical results for, 106.
Brandes, Centripetal Theory of Storms, 163.
Brazil, effect of, on Trade-wind, 71.
Brussels, anemometrical results for, 103.
Buch, Leopold von, on Return Trade-wind, 39.
" Bull's Eye " in storms, 203.

C APPER on Monsoons, 45.
  Carlsruhe, anemometrical results for, 96.
Centripetal Theory of Storms, 163.
Change of Direction of Vane, causes of, 4.
Chiswick, Barometrical Table for, 124.
— — — — — monthly, 133.
— Thermometrical Table for, 135.
Churruca, Don Cosme, notice of Law of Gyration, 108.
Coast-winds of Africa, 67.
Coffin, Anemometrical Tables, 26.
Conclusions, General, on Storms, 274.
Coseguina, fall of ashes from, 37.
Counter Trade-wind in Torrid Zone, 36.
Currents, Upper and Under, coexistence of, 151.
Cyclones. *See* Hurricanes, Typhoons.
— centre of, rule for finding, 217.
— course of, in Monsoon Region, 191.
— — — — Trade-wind Zone, 177
— effects of, 158.

Y

CYC

Cyclones, frequency of, Tables, 181, 294.
— lull in centre of, 196.
— management of ships in, 286.
— origin of, 182.
— — — mechanical, 189.
— stationary, apparently, 182.
— in Temperate Zone, 212.

D AMPIER, notice of Coast-winds, 68.
— signs of a Typhoon, 193.
Dantzic, Barometrical Table for, 124.
— — — monthly, 132.
D'Aprèz, limits of Trade-wind Region, 21.
Dorpat, Barometrical Table for, 125.
Drake, notice of Law of Gyration, 94.
Drift of a ship, affects her position in a Cyclone, 292.
Drury, Byron, notice of Law of Gyration, 114
Duden, notice of Law of Gyration, 93.
D'Urville, Dumont, notice of Law of Gyration, 112.

E ARTH, rotation of, effect on wind, 7.
Earthquakes are independent of barometer, 161.
Espy, views of Cyclones, 172, 230.
Eye of a storm, 282.

F ORSTER, notice of Law of Gyration, 108.

G ALLE, M., Barometrical Table, S. Hemisphere, 134.
Gentil, notice of Law of Gyration, 108.
Glass, on Coast-winds of Africa, 67.
Gnadenfeld, anemometrical results for, 97.
Gogh, Van, notice of Law of Gyration, 115.
Greenwich, anemometrical results for, 101.
Guericke, Otto. von, notice of connection between fall of barometer and storms, 157.
Gyration, Laws of, 11.
— — — proofs of, N. Hemisphere, 87.
— — — — — S. Hemisphere, 107.
— — — recapitulation of authorities, 116

LEI

H ADLEY on return Trade-wind, 38.
Hail, Theory of Formation of, 309.
Hall, Basil, on Trade-winds, 22.
— — notice of Law of Gyration, 112.
Halle, Barometrical Tables for, 124.
— tension of aqueous vapour at, Table, 137.
— — — dry air at, Table, 138.
— Thermometrical Table for, 135.
Halley on Monsoons, 43.
— on Return Trade-wind, 38.
Hawaii, Return Trade at, 41.
Hemisphere, N., rarefaction in, 61.
— S., rarefaction in, 71.
Heywood, notice of Law of Gyration, 111.
Hildreth, notice of Law of Gyration, 93.
Horner, notice of Law of Gyration, 114.
Horsburgh on Coast-winds, 68.
— — Trade-wind, 22.
— notice of Law of Gyration, 109.
Hurricanes.   See Cyclones, Typhoons.
— West India, course of, 176.
Hurricane of December 25, 1821, 162.
— — St. Thomas, 1837, 196.
— — Barbadoes, 1831, 205.
— — Lyons, 1846, 219.
— the Great, of 1780, 209.

I NSTRUMENTS, Meteorological, laws of fluctuations of, 118.
— — diagrams of, 120.

J ANSEN, Lieut., on Monsoons, 46.
— on Barometer in S. Hemisphere, 134.

K ANE, Dr., notice of Law of Gyration, 94.
Kant, notice of Law of Gyration, 90.
Kerhallet on Trade-winds, 23.
Kharkov, anemometrical results for, 104.
King and Fitzroy, notices of Law of Gyration, 110.

L AMPADIUS, notice of Law of Gyration, 90.
Land-breezes, origin of, 5.
Lartigue, Capt., Observations on Winds, 146.
Law of Gyration, 11.
Leichhardt, notice of Law of Gyration, 114

LEN

Lenzburg, anemometrical results for, 97.

Liverpool, anemometrical results for, 101.

London, Barometrical Table for, 124.

— tension of Aqueous-Vapour, Table, 137.

— — — Dry-Air, Table, 138.

Luz, Observations at Gunzenhausen, 130.

MACONET, Dupuis de, notice of Law of Gyration, 94.

Madrid, anemometrical results for, 104.

Marriotte, notice of Law of Gyration, 89.

Maury, Trade-wind Tables, Atlantic Ocean, 25, 29.

— — — — Pacific Ocean, 30.

— Theory of the General Movements of the Atmosphere, 271.

Maxima and Minima (Meteorological), simultaneous occurrence of, in distinct districts, 166.

Melbourne, anemometrical results for, 115.

Meteorological Tables, Northern Hemisphere, 52.

— — Southern Hemisphere, 265.

Moisture, condensation of, effect on meteorological instruments, 138.

Monsoons, causes of, 64.

— duration and changes of, 46.

— early notices of, 43.

— West, of the line in Guinea, 65.

— — — — — in Pacific, 66.

— Region of, Wind-Tables for, 48.

OGDENSBURGH, Meteorological Tables for, 128.

Oxford, anemometrical results for, 103.

PARIS, Barometrical Table for, 124.

— — — monthly, 132.

— Thermometrical Table for, 135.

Piddington, on Mechanical Origin of a Cyclone, 189.

Pliny, notice of Law of Gyration, 88.

Poitevin, notice of Law of Gyration, 90.

Pressure. *See* Barometer, Aqueous Vapour, Dry Air, &c.

RAREFACTION, in Asia, 61.

— — Mediterranean, 63.

— — Southern Hemisphere, 71.

TEN

Redfield and Reid, their Theories of Cyclones, 174.

Return Trade-wind in Torrid Zone, 36.

Romme, notice of Law of Gyration, 90.

Rotations, direct and retrograde, definitions of, 15.

Rotatory theory of storms. *See* Cyclones, &c.

Rules, practical, Frigid Zone, 314.

— — Monsoon Region, 297.

— — North Temperate Zone, 300.

— — South Temperate Zone, 313.

— — Trade-wind Region, 285.

SCHÜBLER, notice of Law of Gyration, 93.

Sea, observations made at, remarks on, 144.

Sea-breeze, cause of, 5.

Seller, limits of Trade-wind, 21.

Ship, motion of, effect on observations of Law of Gyration, 144.

— management of, in Cyclones, 286.

Siberia, notice of Law of Gyration in, 93.

Smith, Piazzi, on Return Trade-wind, 40.

Storm, eye of, 282.

Storm of January 1818, 157, 245.

— — February 1823, 248.

— — January 1850, 222.

— — December 1850, 232.

— — January 1855, 258.

— — Winter of 1855-6, 234.

Storms arising from conflict of currents, 221.

— — — lateral interference of currents, 256.

— retrogression of, apparent, 203.

— rotatory. *See* Cyclones, &c.

— theories of origin of, 262.

— Thermometer sometimes indicates, 245.

Strelecki, notice of Law of Gyration, 114.

Sturm, notice of Law of Gyration, 89.

Sun, rotation 'with' and 'against,' defined, 15.

Swinden, Van, Meteorological Observations of, 130.

TEMPERATURE, Mean, of Torrid Zone, 21.

Teneriffe, Return Trade at, 39.

THE

Theophrastus, notice of Law of Gyra-
    tion, 88.
Thermometer, law of variation of, 135.
Thouars, Dupetit, notice of Law of
    Gyration, 111.
Toaldo, notice of Law of Gyration, 89.
Toronto, Meteorological Tables for, 129.
Torrid Zone, winds of, general review of,
    73.
Trade-winds, deflection of, lateral, 67.
— — — direction of, at edge, 22.
— — displacement in latitude, annual,
    20.
— — effect of Australia on, 73.
— — — — Brazil on, 73.
— — extent and direction, in Atlantic
    Ocean, 25.
— — — — — — Pacific, 30.
— — — — — — Caribbean Sea, 31.
— — origin of, 18.
— — Return in Torrid Zone, 36.
Trade-wind-Zone, Table of Atmo-
    spherical Pressure, 42.
Trombs, Rotation of, 278.
Typhoons, 190.
— course of, 192.
— signs of, Dampier, 193.
— — — Varenius, 194.

ULLOA, Don, notice of Law of
    Gyration, 108.

ZEC

Urbain, Von Wüllersdorf, notice of Law
    of Gyration, 115.

VANE, in what cases its direction does
    not change, 4.
— — — — — — does change, 5.
Varenius, signs of a Typhoon, 194.
Vincent, St., fall of ashes from, 37.

WAVES, atmospherical, 266.
    — Motion of, in storms at sea, 296.
Weather, signs of, Temperate Zone, 302.
Wendt, notice of Law of Gyration, 107.
Wilks, Capt., on Peruvian Monsoon,
    70.
— — on West Monsoons of the line,
    66.
Wind, effect of rotation of earth on, 7.
Wind-Tables.   See   Anemometrical
    Tables.
Wind, NW., prevalence of, in Ireland,
    82.
— SE., prevalence of, in South of Russia,
    82.
Wrangel, notice of Law of Gyration, 93.

ZECHEN, Barometrical Tables for,
    124.
— Thermometrical Tables for, 135.